Fresh Air
FOR LIFE

how to win YOUR unseen war *against*

Indoor Air Pollution

Other Books by Allan C. Somersall, PhD, MD

The Healing Power *of* ^{8}SugarS

What Doctors want YOU to know about Glyconutrients ...
The 8 Sugars Vital to Your Health

The Enzyme Diet™ Solution

How to use the Power of Enzymes
to obtain the healthier body YOU desire!

Nature's Goldmine

Harvesting MIRACLE INGREDIENTS from MILK

Breakthrough in Cell-Defense

How to benefit from the REAL Glutathione Revolution

Your Evolution to YES!

How to become more Positive, Passionate & Productive ...

Understanding The Evolution of YES!

Insights for Affirmative Living

Evolutionary Tales by Dr. YES!

A prescription of Contemporary Stories for Affirmative Living

A Passion For Living

The Art of Real Success

Your Very Good Health

101 Healthy Lifestyle Choices

From The Bestselling Author of ***The Enzyme Diet® Solution***

Fresh Air
FOR LIFE®

how to win YOUR unseen war *against*

Indoor Air Pollution

Allan C. Somersall, PhD, MD

Foreword by Prof. James Marsden, PhD

The Natural Wellness Group
the best of science & nature

Fresh Air FOR LIFE®,
by Allan C. Somersall, PhD, MD

Published by The Natural Wellness Group
A Division of GE Publishers, Inc.
2-3415 Dixie Road, Suite: 538
Mississauga, Ontario L4Y 2B1
E-mail: freshairforlife@aol.com
www.thenaturalwellnessgroup.com

ISBN 0-9737317-1-0
978-0-9737317-1-2

Library and Archives Canada Cataloguing in Publication
Somersall, Allan C. (Allan C.),
Fresh Air FOR LIFE : how to win your unseen war *against* Indoor Air Pollution / Allan Somersall.

Includes bibliographical references and index.
ISBN 0-9737317-1-0

1. Indoor air pollution--Health aspects. 2. Indoor air pollution--Prevention. I. Natural Wellness Group II. Title.

RA577.5.S67 2006 613'.5 C2005-907485-X

10 9 8 7 6 5 4 3 2

Every effort was made to secure complete information on all sources.

Illustration drawings of symbols by Brad Hofbauer

Printed in Canada

This book is dedicated

to

all my teachers

who

first encouraged my scientific curiosity,

or

later sharpened my focus

on health, wellness and the environment.

Disclaimer

This book is designed for educational purposes only. Although the author and publisher are fully committed to providing credible and useful information to the general public, nothing presented in this text should be considered as either medical advice or endorsement for any commercial product. Indoor air quality is determined by a wide variety of factors and each individual will have different sensitivities to various chemical or physical components.

Therefore, the principles presented here should empower the reader to make careful assessment and informed choices to optimize the indoor air quality in any given situation or environment. But if any doubt remains, consultation with a professional of your choosing is always to be preferred.

Acknowledgements

This book is a result of an intense literature search. For that I am indebted to my agent, Carolina Loren, who spent countless hours searching out valuable information for me, and who also guided this project through all its phases right up to eventual completion. Without her diligent support and invaluable skills, this effort would have proven intractable to the author. With such a proficient and dependable agent, writing becomes a true pleasure and not a burden.

I am particularly honored to have Professor James Marsden write the foreword to the book. Dr. Marsden is a distinguished professor and an expert with a worldwide reputation in his field. He has broad experience, coupled with a passion about preserving the integrity of the air, food and water supply for this and subsequent generations. His perspective truly enhances the relevance of the message presented here.

Technically, the entire field of indoor air quality was recently presented in a comprehensive review (*Indoor Air Quality Handbook*, Edit by John D. Spengler, Jonathan M. Samey and John F. McCarthy, McGraw Hill Publishers, 2001). This 1500+ page treatise is such an excellent summary of data for the professional, that one could not but refer to it extensively in writing a monograph for the average reader. I am therefore much indebted to the authors of this major secondary source in specific areas. We have tried to make citation where obviously indicated.

I would also like to express my gratitude to all my editors - they know who they are. Each edit has only improved the quality of the text and the readers will now enjoy the benefit of the final easy-to-read style, while holding rigorously to the content. Special thanks to my wife Virginia, who is the best editor that I have known, with an eye for nuance that other writers would envy.

We are also indebted to The Boomerang Group, Inc. for the innovative cover design, to Brad Hofbauer for sharp illustrative drawings, and to Angela Nurse for the final creative idea on the cover, that gave it the pizzazz it needed.

In the end, all the shortcomings are mine.

ACS, 2006

The earth dries up and withers,
the world languishes and withers;
the heavens languish together with the earth.
The earth lies polluted under its inhabitants;
for they have transgressed laws, violated the statutes
and broken the everlasting covenant.
Therefore a curse devours the earth,
and its inhabitants suffer for their guilt.

-the prophet Isaiah (c.760-690 B.C.)
Isaiah 24:4-6
New Revised Standard Version of the Bible

Table of Contents

Foreword

James L. Marsden, PhD,

The basic necessities of life include air, water and food. Ideally, the air we breathe, the water we drink and the food we eat should all be clean and free of biological, chemical and physical hazards. Historically, unsafe food and water have been major causes of illnesses and death in human populations. However, the twentieth century brought vast technological improvements to address the safety and quality of these vital necessities.

Today, water is systematically purified by municipal water treatment systems and is often further treated by consumers or manufacturers of bottled water, in order to make it as safe as possible. For the same reason, food products are often heat pasteurized or sterilized to eliminate harmful bacteria. Even fresh, unprocessed foods are being harvested in ways that reduce the presence of biological and chemical hazards.

The fundamental safety of both food and water has been dramatically improved through the widespread application of different technologies. These achievements, in combination with advances in sanitization, have greatly contributed to the increases in life expectancy that have occurred over the past century and a half. In 1850, for example, the life expectancy of an infant was estimated to be 38.3 years. Then in 1900, life expectancy was 48.2 years. But lately, in 2005, the U.S. Dept. of Health and Human Services estimated that the life expectancy of an infant born in the United States has increased to 77.6 years. Yet there are still many challenges involving the safety of food and water, and opportunities for improvement in reducing cases of food and waterborne disease.

In contrast to the advances in the safety of food and water, **the safety and quality of the air we breathe seem to be moving in the wrong direction**. Contamination of indoor air and outdoor air is well documented and is a major area of national concern. The reasons for deterioration in air quality are many and diverse. Certainly, the industrial age and the advent of the automobile have had an impact on contaminants in the air. In this age of air travel, biological hazards are spread much more rapidly from one part of the world to another. Natural phenomena like the eruption of Mount St. Helens also have an enormous effect on particulates in the air.

Fortunately for mankind and the planet earth, the sun has a cleansing effect on the outside air. Particulates are ionized by the sun's energy and fall to earth. The ultraviolet energy from the sun also destroys biological and chemical hazards. Yet in large cities, man's ability to create air pollutants sometimes exceeds the ability of the sun to clean the air. Therefore, in the 1960's, air pollution emerged as an important environmental issue. As a result of intense efforts on the part of industry, government and individuals, and the forces of nature, outdoor air quality is improving. But clearly, our nation and the world as a whole have a great deal left to do to address the problems of soot, smog and excess ozone levels. Outdoor air pollution is still a major cause of human illness and some have predicted that **unless we improve the safety and quality of the air we breathe, the improvements in life expectancy may be reversed and we may see real declines in the next century.**

The safety and quality of indoor air represents another facet to the problem of air pollution. In response to the energy crisis that occurred in the 1970's (and which is still in full force today), our homes, schools and office buildings have been basically sealed. Pollutants that find their way into the indoor environment often stay there. Gone are the days when we allowed fresh air and sunlight to purify our indoor environments.

In this book, **Fresh Air for Life**, Dr. Somersall addresses the issue of indoor air safety and quality. He identifies the origins of biological, chemical and physical hazards that affect indoor air quality and makes the case for relevance by correctly pointing out that we spend about 90% of our time indoors. Dr. Somersall identifies indoor air pollutants which are difficult or impossible to avoid. These include chemical (carcinogens), biological (viruses and other microbes) and physical hazards (particulates and fibers). The sources of these hazards are also many and varied, and might include, for example, pets, plants or puffs of second hand smoke. They are ubiquitous in our world and make up at least what the author calls the "Ten Contaminants you can't avoid". The public health consequences of these contaminants are also described in some detail in the book. These include cancer and infectious diseases, ranging from influenza and the common cold, to Legionnaire's disease and SARS.

Most importantly, rather than highlight the dire consequences of indoor air pollution and leave his readers in a state of fear and helplessness, Dr. Somersall also identifies practical steps which can be readily taken to mitigate risks. He explains action strategies that will effectively improve indoor air quality and reduce the risks it would otherwise pose to you and your family. In addition, he discusses state-of-the-art technologies that are available to consumers for purifying indoor air. And he writes for the consumer in terms that you can understand. That is welcome at a time when the public is so unaware or else misinformed about indoor air.

As a scientist, I strongly recommend this book for its excellent treatment on the subject of indoor and outdoor air pollution. However, I recognize that most of the readers will not be scientists or medical doctors, and just as well. More likely, you are a parent, a grandparent or simply an individual concerned about creating a healthy environment for you and your family. **This book has the most to offer for a person just like you.** It explains the problems and the potential dangers associated with airborne contaminants, and more importantly, it offers you some solutions to those problems.

Whether or not you are now aware of the state of your indoor air environment, I am sure that you will benefit from **Fresh Air for Life** as you come to terms with the real implications of this important pollution challenge. You will learn how to win your personal battle against Indoor Air Pollution and your family will realize significant health benefits as you implement the strategies so carefully elaborated in this book.

As you read, I hope you will enjoy this breath of fresh air, for a change!

Dr. James L Marsden,
Regent's Distinguished Professor,
Food Safety & Security,
Kansas State University

Preface

This book is long overdue. It is a book about a truly indispensable necessity of life - Air! It's about **Fresh Air** - finding real fresh air in the modern polluted indoor environment where you may have become a prisoner of your own neglect. As such, it is also about You - about the quality of your health and life indoors, and perhaps what you might still be taking for granted. You have more than likely sealed yourself in with contaminants that you cannot see and these invisible enemies are defeating you all the time. It's a losing indoor air pollution battle of which you may even be unaware. But denial is no longer an option.

As a writer on health and wellness, as a physician and scientist, and as an international speaker on lifestyle issues for more than two decades, I can't believe that this important issue of indoor air pollution passed right under my own radar screen. But then, I must be so typical of the population at large - oblivious to something so immediate and intimate, it's ridiculous!

However, once I started doing the research for this book, I was alarmed. You will soon begin to learn what I found out and why it has given me a new passion and mission. Everyone needs to read this book. You will become aware and alert to the indoor hazards that surround you all the time, and these are things you can do something about if you choose to.

If you have an open mind and you are willing to learn, you will find it hard to put this book down. You will soon discover some things that you will wish you had known a long time ago.

This book is not designed to curse the indoor darkness - not at all. Rather, it will light a candle of hope. It will offer alternatives to help you enjoy **Fresh Air for Life** as you learn how to win your personal battle against Indoor Air Pollution.

The book is divided into **three equally important parts**.

In Part One, we go to great lengths to detail the exact nature of the **Problem** with indoor air pollution. It is indeed a problem that is not readily apparent, and no thanks to the media, it has been obscured and overshadowed by the prominent exposé of visible outdoor pollution. That outdoor focus has been pivotal to the activists in the early environmental movement and makes for good television. But it is misdirected.

The major problem indoors has only become worse. There are real and immediate *concerns* that you cannot ignore because there are many *contaminants* that you cannot avoid. If you do, you are likely to suffer health *consequences* that you cannot afford. With the apparent epidemic of asthma and other respiratory conditions, the recent SARS outbreak, the re-emergence of some infectious diseases, the threat of a potential worldwide pandemic, and the horrid nightmare of bioterrorism - all this and more has awakened people everywhere to the importance of the air they breathe and the value of clean, uncontaminated **Fresh Air for Life.**

But there remains so much ignorance and misinformation regarding indoor air pollution!

The U.S. EPA has already concluded that indoor air is often far more polluted than outdoor air, so the focus has truly been misplaced for quite some time. Since we spend 90% of our time indoors, confined in 'tight' contaminated buildings and polluted automobiles, this problem cannot be neglected any longer. That's why I've written this book.

Every responsible writer on Wellness issues faces the challenge of making the credible case for health responsibility and wise lifestyle choices leading to behavior modification, *without* resorting to sensationalism or needless exaggeration. It is true that human nature resists personal change that costs in either comfort or convenience. It is also true that modern culture is gullible and suffers from information overload to the point of mental numbness. But that does not justify manipulating the real information in order to make a more compelling case. Rather, the facts alone must always speak for themselves, as they do here. Each individual must be receptive to new credible information and provocative ideas, then assert their own values and take responsibility in return.

Therefore, you must judge the facts about indoor air quality and make the appropriate response to this problem, for your own improved well-being. The information in this book will give you knowledge, and therefore the opportunity, to make healthy choices for a cleaner indoor environment and so enhance the quality of life for you and your family.

In Part Two, we propose a comprehensive **Solution** to the indoor air pollution problem based on ***Five Essential Strategies***. We first deal with the *three traditional methods* of cleaning the air: (1) reducing or eliminating *sources* of pollution; (2) increasing *ventilation* to dilute or replace the polluted air, and (3) *filtering* out any suspended particles, bioaerosols and even volatile gases, with mechanical and adsorption filters.

But those conservative approaches, even when taken together, could never be the final answer indoors. The fourth strategy therefore goes on to examine the *three conventional methods* employed in nature outdoors: *ozonation, ionization* and *UV radiation.* Yet each of these also has its own limitations, as well as advantages. The ultimate solution is therefore reserved for a space-age technology that actually derives from a solution to NASA's problem with indoor air.

The third generation of this technology known as *Radiant Catalytic Ionization*™, proves that it can deliver a complete answer for purifying the indoor air. This final strategy exploits ***Five Important Features*** that together provide a synergy that combines advanced oxidation and plasma chemistry to get the job done. What is more, the science supports it and the test results will speak for themselves. ***It is the only space-certified air purification technology anywhere in the world today***.

In Part Three of the Book, we make the **Application** of this relatively new technology for indoor air purification to your ***Five Spheres of Activity*** which cover increasing dimensions of modern life. We choose to begin with your *personal* space, with an emphasis on a personal air purifier that effectively forms an electrostatic shield to protect your breathing zone. Then we focus on maintaining your *home* space where you spend so much of your time. A clean indoor environment with purified **Fresh Air for Life** will make a real difference for you and your family. But since you can't be imprisoned at home all the time, we look next at protecting your *traveling* space.

North Americans spend perhaps too much of their time confined behind the wheel of automobiles, in tunnels of unseen exhaust pollution, and in airplane confines, breathing re-circulated air possibly contaminated with who knows what. We broaden the sphere further by examining the issues at *work* where you also spend a lot of your time and where you can have influence.

Finally, we expose the challenges in other *commercial* venues such as health-care facilities, schools, hotels, etc. In all these Spheres of Activity where you live and work, the practical applications of this technology have proven to be convenient and most effective.

The published literature in the field of indoor air quality is vast and therefore overwhelming. But it remains obscure and in the professional domain. Therefore, I wrote this book with a person just like you in mind - not the skilled technician or professional, but the regular person going about your daily life, typically oblivious to what is at stake in your immediate indoor environment. You're not alone. I've tried to write in common language as much as possible and to keep things simple enough, without doing disservice to the facts. That's been a real challenge. However, you will appreciate that to make the subject matter more user-friendly, most of the key ideas are highlighted in one way or another. Whatever else you do, as you read this book, extract the useful and practical highlights and put them to work for you.

Poor indoor air quality is the big environmental secret of our times. But it's time for this book to blow the whistle and to let the secret out, so that those who care about their health and quality of life can take advantage of the latest space-age technology to remedy the situation. You will soon join their ranks as you make some changes and apply the comprehensive strategies you are about to learn. These will allow you to enjoy **Fresh Air for Life** and so win your personal battle against the silent enemy that now demands your attention: INDOOR AIR POLLUTION!

You will win!

Dr. Allan C. Somersall,
Toronto, Ontario

Part I

The PROBLEM

Untold thousands are dying annually from Indoor Air Pollution

ONE

Introduction

"We've had our priorities wrong, ***it's an indoor problem.****"*

Life is invaluable and even at its best, it remains a story of **survival**. To be sure, your own survival critically depends on your immediate local **environment**. That environment is exactly the very **air** you breathe.

Just think of it. Fish survive in water, and only in water. From the tiny goldfish in a little kid's hobby aquarium, to the great monster whales that dominate the ocean depths, they all thrive in water as their natural habitat. Similarly, a great diversity of birds traverse the open skies with grace and splendor - sometimes alone, in delicate curiosity, or else in organized flocks in search of safety and adventure.

But what of human kind?

With our feet planted upon *terra firma*, we naturally inhabit the atmosphere. We sustain our life by constant interaction with our ethereal surroundings. **The environment of AIR is our true habitat, our indispensable home**. In it we thrive, we live, we move and maintain our very existence. Yes, we can dive into the sea, or take our feet off the earth and even dare to fly, but we cannot disengage for more than just a few minutes from the very air we breathe. It is indeed the breath of life! The *sine qua non*! (Without this … nothing!).

When you think of it, you might say 'that's so obvious'. Of course, we must have air! But it's everywhere - so invisible and so silent - that we take it all for granted. If it were not for the wind, we might even wonder if it was really there at all … until we tried to hold our breath, and thereby foolishly threaten our own demise. **Constant breathing is our most intimate interaction of any kind**. It is involuntary and innocuous, but also indispensable.

Just to emphasize how intimate this interaction with our air environment truly is, consider the volume of air you must process 24 hours a day, seven days a week. There's nothing else like it. Typically at rest, you take about 15 breaths per minute. In each one, you inhale about 500ml or so of air. That amounts to 7.5 litres per minute, or 450 litres per hour, or more than 10,000 litres (400 cubic feet) of air every single day. That's what you breathe in and breathe out. It's an exchange you can't avoid: the 'breath of life' that maintains 'the living soul'. Would that it were nothing but Fresh Air!

Fresh Air for Life!

THE IMPORTANCE OF AIR

In any critical situation, from simple first-aid to life-and-death trauma, professionals have all learned the **ABC** of urgent care. It's a paradigm that must always be the first consideration. '**A**' is for the all-important **Airway** - you must have some access to get the all-important air into your body. '**B**' is for essential **Breathing** - if you can't do that on your own, then it must be done for you, mechanically. Finally, '**C**' is for **Circulation** - the blood must take the vital oxygen derived from the air in your lungs to every living cell in your body. It then returns with carbon dioxide for expiration. That's basic physiology, but what a life-saving and life-sustaining exchange activity!

But what are we exchanging in the exercise? We're getting more from the air today than the oxygen that we need. No one in industrialized society these days would deny that **air pollution** has become a real problem - not with the common sight of billowing smoke from burning fossil fuels in industrial plants or factories, and all the other widespread symbols of modern industrialization. Or think of the dense smog and the summer haze that so often covers Los Angeles, Houston or any other large metropolitan city like a blanket. And then there is the constant trail of exhaust smoke that automobiles spit out in congested commuter traffic every day. These pictures are self-explanatory.

This is not a new phenomenon, but it is a relatively recent development. Our great grandparents lived much simpler lives - closer to nature and much more in harmony.

Widespread concern for the endangered environment, plus increasing public awareness of air pollution in particular, and the consumer obsession with smog levels, all took hold with the Baby Boomers' generation. In the latter half of the twentieth century, as city sprawl became the norm in the industrialized west, air pollution became a rallying cry for social and political activists, as well as conservationists and the media, thereby arousing public sensitivity and reaction.

Despite our good intentions, however, **we seemed to have had our priorities misplaced**. We were quick to acknowledge the problem that we saw with our naked eyes and pushed for legislation and technology to relieve the contamination of the precious air we breathe *outdoors*. But in reality, we do spend most of our time *indoors*. Yet the early assumption was that if we could 'clean up our act' outside, we could return to our 'safe harbor' inside and shut the door. Then and there, we might enjoy the best of health and wellness which we all deserved.

But it just wasn't so! It soon became apparent when the first measurements of indoor air quality were made, that there was no less significant a problem inside homes and other buildings, as there was outdoors. Since then, the history and evolution of research on indoor air has closely intertwined with investigation of outdoor air pollution. The methodology and techniques used in either case have been comparable and the results and health consequences of personal exposure have *both* proved alarming in their own right. They raise some major concerns you can't ignore.

IT IS AN '*INDOOR*' PROBLEM

The stark reality is that we North Americans spend the vast majority of our lives inside buildings and automobiles. For better or for worse, there has been a dramatic lifestyle shift in the last century. The 1930 US Census data showed that as much as 80% of Americans lived on farms at that time. But by 1950, a mere twenty years later, 80% of the population was now residing in cities. This large and rapid shift to urbanization meant that we changed much of our working environment from fields to floors. We moved from outdoors to indoors, with major implications. We turned away from the natural environment to the artificial, machine-operated, fossil-fuel-run, working environments to which we commuted, again in fossil-fuel-powered automobiles, trains and airplanes. **Today, people in industrialized countries spend more than 90 percent of their lives inside buildings,**(1) **equivalent to over 65 years of the average life.**

For infants, the elderly and persons with chronic diseases, that percentage of time indoors may be even higher. These groups tend to be less mobile, show more dependence on others and generally have less reason to be outside. Full-time homemakers and small children may spend over 21 hours per day inside the home, with another 2.5 hours inside other buildings like stores or offices, or in transit vehicles. Most urban residents of all ages also work indoors and enjoy lifestyles that trap them in walled surroundings at home and in their communities. Even when employed, these city dwellers spend about the same

90% or more of their time indoors, 63% of it at home and 28% at work.(2) And then, when we're not in buildings, we spend more time trapped in our cars than we do running, walking, standing or just sitting outside.

Children are especially vulnerable and deserve a further comment or two. They are more vulnerable to toxic vapors simply because of their higher metabolic rates. They breathe in more than twice as much oxygen - and therefore, toxins too - relative to body size, when compared to adults. They tend to be more active, which increases their breathing rate and for smaller children especially, they usually play close to the floor where the heavier pollutants settle. The more modern buildings where children go to school are designed today to conserve energy and therefore are more 'weatherized' - they have reduced air exchange with the outdoors.

Today, people in industrialized countries spend more than 90% of their lives inside buildings, equivalent to over 65 years of the average life

In fact, the design of most of the modern buildings in which we live and work has changed in recent times. The energy crisis of the 1970's provided incentives to conserve fuel. To increase building energy efficiency, **the architects and builders designed these more air-tight buildings that now enclose a Pandora's box of airborne toxins that have nowhere to go but into their inhabitants' lungs.** That's you and me.

Researchers at the Lawrence Berkeley Laboratory (LBL) Indoor Environment Program - the nation's oldest indoor air research program - have been investigating the relationships between indoor air quality, energy use in buildings, and the health and comfort of the people who inhabit these spaces. That program grew out of concern about the impact of energy conservation on buildings.(3) The program leader and chemist, Joan Daisey, put it simply and directly, "*if you tighten up buildings to save energy, you can adversely affect indoor air quality.*" She was so right.

Clearly, you can trace some of the problems of indoor home pollution to a noble effort - energy efficiency. But those good intentions have produced some adverse consequences. "We're building tighter and tighter homes and bringing all sorts of products and chemicals into them, but there's less and less air going through," said Richard Corsi, director at the Texas Institute for the Indoor Environment at the University of Texas in Austin. **"A tighter home is good from the standpoint of conserving energy but bad in terms of occu-**

pants inside."

A recent study found that the allergen level in 'super-insulated' homes is 200% higher than in the more 'ordinary' homes. It's no wonder we have to endure such insults, when we realize that nature's air-cleansing agents are kept on the *outside*, while we contend with the consequences of air pollution on the *inside*.

According to a December '98 article in *The Wall Street Journal*, "Carpeting, poorly ventilated fireplaces, mold, bacterial toxins, dust mites .... **an almost endless collection of highly allergenic products has invaded our homes and we have sealed them in with deadly precision**."(4) Should the consequences of this not be called '*tight building syndrome*'? It often is.

The presence of an increasingly complex array of synthetic chemicals used in building and insulating products, furnishings, carpets, consumer products and hobby and craft materials further compounds the dilemma. Thousands and thousands of new molecules, limited only by our human imagination and creativity, have become part of our daily lives. Even the use of traditional combustion fuels like wood, kerosene and coal, as well as the use of alternative heating and cooking systems, if not properly designed, installed and maintained, can release combustion by-products into the air that prove very harmful to health. Indoor exposure to contaminants originates not only from the myriad of indoor materials and combustion products, but also from the normal dead skin flakes shed by humans every day, plus their pets' hair and dander, and of course from personal activities such as smoking in the home. Pogo, the comic strip character, put it succinctly into these words that summarize the indoor air pollution epidemic: "*We have met the enemy, and he is us.*"

There is a growing body of scientific evidence to indicate that the air in homes and other buildings can be more seriously polluted than the outdoor air in even the largest and most industrialized cities. In fact, based on a 1987 comparative study of environmental problems in the U.S., **the *Environmental Protection Agency* (EPA) ranked indoor air pollution among the *top five* most significant environmental dangers to the American public**.(5) Their research concluded that indoor air can be two to ten times (and sometimes as much as 100 times!) more polluted than outside air!(1)

According to the American College of Allergists, **50% of all illnesses are either caused or aggravated by polluted indoor air.**(6) The National Academy of Science has estimated that 15% of Americans suffer symptoms from indoor air pollution but others have put that figure even higher. One official with the EPA suggested that it is as high as 40% .(7) A subsequent study in the *American Journal of Public health* (2002) estimated that each week, 35-60 million U.S. workers have symptoms of illness related to their workplace. That is at least 40% of the 90 million non-industrial indoor workers.(8)

But it is still popular in many circles (thanks to the media!), to think that it is the air outdoors that presents us with the greatest risk. However, after more than two decades of solid research, all the facts now point in the opposite direction. It is actually the air inside our homes, schools, offices, hospitals, factories and other buildings, that is truly the most harmful.

There is now a virtual indoor air pollution epidemic!

Many studies have been done on human exposure to environmental pollution. In the absence of any major tragedy or natural disaster, almost all of them prove to be consistent with a general rule for the concentrations of volatile organic chemicals, namely -

Personal Exposures > Indoor Concentrations > Outdoor Concentrations

and often in a similar ratio of about 3:2:1.(9) The major sources of exposure to all chemical groups studied have been relatively few in number and tended to be physically close to the person - usually inside that individual's home.

Nothing could be more at odds with the conventional wisdom which seems to suggest that the major environmental pollution is outdoors. That perspective which focuses on the conditions outside, would seem to make sense when we are so frequently and consciously enveloped by smog and choked by filthy exhaust from automobiles. It may be actually true for a fraction of the little time we do spend outside, inhaling the ambient air. So the popular media would have you believe that the major sources of pollution are indeed the reckless wheels of industry, gas-guzzling automobiles, urban sewage treatment plants, toxic landfills and hazardous waste sites.

But again, it just isn't so! That's not where the major problem lies. The population at risk is really the general population, confined by lifestyle choice (and sometimes by necessity) to inhabit for the vast majority of the time, closed and relatively air-tight buildings, and the transport vehicles that shuffle us between them. **We essentially live and work indoors in an environment that is far from healthy,** as we shall soon see.

The fact that we spend more than 90 percent of our time indoors was first reported by the EPA but was also confirmed in a May 1999 national survey conducted by the American Lung Association (ALA). Ironically however, in that same survey, again 90 percent of homeowners questioned were not aware that poor indoor air quality could be a problem. And therein lies the dilemma. We have a problem and don't yet know it. But all is not lost, for in another nationwide survey, about 95 percent of people who responded said they thought the quality of air in their homes was either somewhat or very important.(10) And who wouldn't think so? The hope is then, that increased awareness through education will lead to some kind of responsible action.

But don't wait for the professionals who should be responsible. It seems that many policymakers and those in positions of authority have not quite understood or even received the true message. If they did, why would we be spending millions of dollars monitoring the outside air and perhaps many billions trying to improve its quality, when we spend so little time outside exposed to pollutants provided by that direct route? Why would the same U.S. EPA spend ten (10) times as much money on 'clean air *outdoors*' than on clean air *indoors*?[8] Why then would we spend more than twenty (20) times as much on research for outdoor air quality versus that for indoor air? Why then too, is the focus of control on standards for national air quality, industrial production and exhaust emission levels and not on some more innovative approach to human exposure levels? Perhaps we have put 'the cart before the horse' or we have just been 'missing the forest for the trees'!

For the record, in the 1989 EPA Report to Congress on this subject, it was estimated that the aggregate costs from indoor air pollution would amount to tens of billions of dollars per year.[5] The measurable economic impacts which were considered did include: materials and equipment costs, direct medical costs and lost productivity. The published literature contains strong evidence that symptoms occurring in the workplace significantly reduce productivity and severely tax healthcare systems.

Medical costs related to indoor air quality far exceed the costs of sickness tied to outdoor air. Estimated annual costs and productivity losses in the United States reported in 1998, have been put at $1-4 billion related to allergies and asthma, $6-19 billion related to respiratory disease, and $10-20 billion related to symptoms of the so-called 'sick building syndrome'.[11] These are potential savings or productivity gains that can be achieved by improving indoor air quality in the workplace.

And that's not all. Another report in *Business Week* (2000) suggested that U.S. companies could save $58 billion annually by preventing 'sick building syndrome' and an additional $200 billion in worker performance improvements by creating offices with better indoor air.[12] Even U.S. Department of Energy studies suggest that improving building and indoor environments could reduce health-care costs and sick leave and increase worker performance, resulting in an estimated productivity gain of $30-150 billion annually. The California Air Resources Board just reported to the State Legislature (2005) their $45 billion annual estimate of the cost of poor indoor air quality in that state alone. Such cost estimates should sound a national alarm! Yet the silence in this area is deafening.

But the effects are much more than economic. They are personal and directly alter the true quality and even quantity of life per se. *Every day, people suffer and die from the consequences of indoor air pollution.* In the stillness of the air, and the quiet sounds of shallow breathing inside many a building or automobile, we are generally oblivious to the true unhealthy state of affairs that result from the inadvertent contamination of the surrounding environment. However, those are concerns you can't ignore.

In 1989, the *World Health Organization* (WHO) estimated that 30% of all modern buildings had indoor air pollution problems.(13) In 1993, the *National Institute for Occupational Safety and Health* (NIOSH) reported that they received more than 7,000 worker complaints about building-related symptoms.(14) In 1994, the *Organizational Safety and Health Administration* (OSHA) estimated that there were more than 1.3 million buildings with poor indoor air quality in the U.S., potentially affecting more than 21 million employees.(15)

This is serious business. But you need not become paranoid or pessimistic too soon. It would be ridiculous to be walking around your home wearing a gas mask and a blue contamination suit. That would be overkill - far too defensive and far too alarmist. In fact, it would be pointless. What is certain is that the air quality and health of your home and mine, as much as our workplace - be it office, factory or farm - will be a major challenge at the forefront of technology in the coming years.

FINGERS ON THE PULSE

Just to illustrate, let's go to a leading research think-tank such as the Battelle Memorial Institute. This global leader in technology development, product development and commercialization began operations in 1929. It manages and co-manages a staff of 19,000 scientists, engineers and support specialists in 70 locations throughout the world. It serves 1500 commercial and government clients and conducts almost $3 billion in annual R & D, leading usually to some 50-100 patented inventions in the typical year. Battelle counts among its major successes the development of the office copier machine (xerography); pioneering optical digital recording technology that led to the compact disc; medical breakthroughs such as reusable insulin injection pens and other drug delivery technologies; the development of the machine readable barcode symbol (UPC) that is now found on millions of consumer products that are quickly scanned in supermarkets and retail stores; first generation nuclear fuel rods; the US sandwich-mint coin, and advanced battery systems utilized in the B-1B Bombers. That's just to name a few.

Now, here's the point.

For nearly two decades, Battelle has investigated and researched indoor air contaminants.(16) The focus has been on pinpointing emission sources, developing emission reduction approaches, monitoring personal exposure, and detecting exposure pathways. For example, Battelle's personal exposure monitoring has included studies of the pathways by which children are exposed to pesticides and organic chemicals in house dust, food and air. Monitoring of human breath composition has been used to document exposures to chlorinated organic chemicals in the workplace and at home.

The Environmental Protection Agency (EPA) ranked indoor air pollution among the top five most significant environmental dangers to the American public

The emission rates and fates of nitrogen oxides and other pollutants emitted from indoor combustion appliances have been studied for the Gas Research Institute. That work elucidated the indoor reactions of nitrogen-oxides and showed that burners designed to minimize nitrogen oxides, also tend to increase emissions of formaldehyde, carbon monoxide and other pollutants. It also identified materials suitable for passively removing nitrogen oxides from indoor air.

Recent studies for the American Petroleum Institute have addressed indoor particulate sources, and the impact that outdoor air pollutants have on *indoor* air quality. In one study, the normal activity of having a shower was shown to be a significant source of fine particles to the interior of an occupied home. Particle production was dependent on the duration and intensity of the shower, and on the temperature and hardness of the water. In another study, penetration of outdoor air pollutants to the indoor air was studied in an occupied residence located near heavily traveled roadways. They monitored the hourly average indoor concentrates of benzene, formaldehyde, carbon monoxide, nitrogen oxides and particulate PAH (polyaromatic hydrocarbons).

Another extensive study to explore the nature of trace organic and inorganic chemicals emitted by natural gas appliances in 'normal' unvented indoor use proved quite revealing. Nitrous acid, formaldehyde, traces of hydrogen cyanides and nitrous oxide were all identified and quantified, as well as ultra fine particles and particle-bound PAH. Some experiments were also performed by deliberately using improper appliance operating conditions, to explore the effect of emissions.

And then, here's the clincher …

A group of Battelle scientists and researchers have examined all the

Healthy Home Trends for this first decade of the twenty first century and pin-pointed the *Top Ten Trends in Healthy Homes for 2010* in order of importance. **After thorough inquiry and assessment, their Number One pick in this health category was Breathing Easy: Indoor Air Quality**!' They too observed that increasingly energy-efficient homes have created interiors that are "a virtual soup of odors and fumes from indoor pollutants." They predicted that what you are likely to see on the market and in homes by the year 2010 are products for advanced air venting, filtration and purification and biosensors that help fight humidity, mold and other indoor pollutants.(17)

To put this in context, **the focus on indoor air quality was put ahead of all other major health innovations and domestic technology in the marketplace today**. These would include the following: home-based medical monitoring diagnosis and care; home safety and security monitoring; reliable electrical power; whole-house water quality; healthy and safe foods; anti-aging (the baby-boomer factor); battling mites and molds; labor-saving cleaning devices and germ-resistant materials. None of these other leading trends proved to be as significant in Battelle's estimation, as the need for concern about indoor air quality.

As Battelle provides solutions and helps develop innovative products for commercial customers by leveraging technology into competitive advantage, they must have their fingers on the pulse. I would bet on that, would you?

Clearly, indoor air pollution raises some major concerns today that you simply can no longer ignore. That's the subject of the next chapter.

TWO

Concerns You Can't Ignore

*"The Air you breathe may be **dangerous to your health.**"*

Air quality relates to people and air pollution does have consequences to human health. Poor indoor air quality has been shown to cause or contribute to the development of chronic respiratory diseases like allergic asthma and hypersensitivity puenmonitis. It is a major contributor to the spread of infectious disease and can cause toxic reactions leading to some forms of cancer. At other times, the symptoms are very diverse and non-specific and no immediate diagnosis becomes apparent. Symptomatically, it causes headaches, dry eyes, nasal congestion, nausea, fatigue and a host of other common immediate complaints. In the extreme, however, indoor air contamination can have some other consequences that later prove to be deadly. These are all concerns you can't ignore.

Some people are much more susceptible than others to indoor air contaminants and allergens. These would include obviously the *asthmatics* who have sensitive airways and react to a wide variety of triggers which cause their airways to constrict. For such patients, this can cause mild to moderate shortness of breath, wheezing, restlessness, tightness in the chest, coughing and the like. Others who are at increased risk of symptoms from poor air quality would be *the elderly*, plus *patients with chronic lung disease*, and *small children* with immature or sensitive soft tissues in the nose, throat and upper airways.

The American Lung Association has been emphasizing that air pollution contributes significantly to lung disease, including respiratory tract infections, asthma and lung cancer.(1) All forms of lung disease taken together, claim almost 335,000 lives in America each year and is the third leading cause of death. In the past decade, the death rate from lung disease has risen faster than for almost any other major disease, despite the decline in smoking and improved medical management of respiratory illness. Now, that should be a surprise. Do you wonder why?

The health consequences of air pollution go beyond the obvious lungs and respiratory system per se. Other organ systems can also be affected. In fact, a remarkable new study published in *Circulation*, a journal of the American Heart Association, suggests that air pollution in U.S. cities may even now cause twice as many deaths from heart disease as it does from lung cancer.(2) In addition, we should now include even diabetics who have also proven to be vulnerable to air pollution,(3) and anyone else who is immune-compromised or simply, very ill.

But none of us is exempt. Virtually everyone is affected in one way or another. And despite the labeling of some 'sick buildings', giving rise to "sick building syndrome", we should consider no home or other building to be immune to this indoor air pollution epidemic. Ironically, **even the very headquarters of the U.S. EPA which was constructed not very long ago, proved to be a source of illness for many EPA employees!**(4) The building was "sick", perhaps both by design (intrinsically so) and by default (contaminants brought in from the outside). This only underscores that if the headquarters of the very government agency charged with monitoring, researching and finding solutions to this serious health problem, could in fact prove to be defenseless - then almost any other home or building could be presumably afflicted in similar fashion.

The EPA used their own criteria to determine that more than half the homes and other buildings in the United States are indeed 'sick', implying that in one way or another, they could be hazardous to the health of occupants as a result of airborne contaminants, irrespective of how clean or even sterile they may appear. Again, it's a concern you can't ignore.

A WIDESPREAD EPIDEMIC

If you think that it's only your family that has seen an increase in breathing and sensitivity problems, you should think again. You are not alone. You're part of an alarming trend. Just consider the following reports which underline the extent of this epidemic.

U.S. News & World Report (May 8, 2000) labeled the recent trend as "An Allergy Explosion." *The Chicago Tribune* (September 26, 2004) published a comprehensive article on the astounding 86% increase in asthma cases between 1980 and 1996. Indeed, asthma cases have more than doubled in the past 20 years. According to the 2001 *National Health Interview* Study reported by the National Center for Health Statistics, over 7.6 million children (5-17 years) and more than 12.7 million young adults (18-44 years) in the U.S. suffer from asthma. Other estimates now show that about 6% of the American population as a whole suffers from asthma. In more general terms, a *Life* magazine cover reported that as many as **50 million Americans suffer from some sort of breathing difficulty.** Because asthma is more prevalent in children, its prevalence is a whopping 11% of all children in the United States. In effect, it is the leading cause of school absenteeism (estimated at 10 million missed school days each year) and recorded pediatric hospital admissions.

USA Today (September 30, 2004) reported that household chemicals are linked to the increasing incidence of childhood asthma. The many compounds in today's building materials leak out of their source products and gradually become part of today's home indoor air contaminants. *The American Academy of Pediatrics* points out that children are more vulnerable to these and other airborne contaminants than adults. This elevated risk derives from the cellular immaturity of children and their ongoing growth process. Even minor irritations caused by air pollution which would produce only a slight response in an adult, can result in a dangerous level of swelling in the lining of a child's narrow airways. Increased exposure to air pollutants during childhood increases the risk of long term damage to a child's lungs.

Nature's Home on-line describes a most interesting scenario. Imagine that in any large city or even in small town America today, a young couple receives the wonderful news that they are expecting their first child. Of course, they're very excited. They prepare a special nursery in their home for the baby's arrival. They paint the walls, put up wallpaper, install new carpeting, hang new drapes and buy new baby furniture. Unfortunately, every one of these materials can emit toxic gases which can cause headaches, dizziness and respiratory symptoms due to nose and lung irritation. This is especially true for the young, fragile and vulnerable infant. If parents like these only knew the real environment they would be taking their precious baby to when they leave the confines of the hospital … just the thought of such a potential toxic environment would make them sick! But that happens every day. And the hospital scene is no better. If they think of the airborne infectious agents alone in that institution, they would still want their new baby out nonetheless.

It is also noteworthy in this regard, that asthma is not the only effect in children. *Oprah Winfrey* dedicated an entire show to the problem of indoor air pollution in public schools and how it may contribute to learning disabilities and other health problems in children. A growing body of scientific evidence has revealed that children who live in homes that are well ventilated, dry, and free of pests, poisons and dangerous gases, will be healthier and lead fuller lives.

Overall hospitalization and doctors' visits due to asthma have shown dramatic increases and despite the recent advances in treatment and management of asthma, death rates due to this disease have tripled since 1976. About 5,600 people actually die each year in America from asthma itself and its complications. However, asthma is a relatively new condition, having developed it would appear, at the turn of the last century. In some countries, where the population still lives on the land, and is removed from 'sick buildings' and toxic emissions from industrialization, asthma is a relatively rare condition. It is more than coincidence that the worsening prevalence of asthma is seen to parallel the worsening of indoor air quality. The rapid rise of both, points to an environmental cause and not to some dramatic change in genetics. And that environment is essentially indoors where we live and work and where the air is now proven to be polluted with allergens and contaminants that might surprise us all.

POLLUTANTS

What might these contaminants be? They are many and varied, and contribute surprisingly but significantly to the nation's state of health. Let's take a preliminary look at some of the major classes.

Biological pollutants (which include molds, bacteria, viruses, pollen, dust mites and animal dander) contribute to poor indoor air quality that leads to breathing problems, allergic reactions, some forms of infectious diseases, and more. Unsuspected *mold* like *stachybotrys chartarum*, have been identified with upper respiratory tract irritation and acute pulmonary hemorrhage in infants. It grows readily on wet cellulose material. Most airborne *bacteria*, may not be pathogenic but, in susceptible individuals, a few strains can give rise to serious infections like Legionnaires' disease or the milder Pontiac Fever. Different airborne *viruses* can cause the 'common cold' or systemic influenza, which are usually mild illnesses. Others give rise to more aggressive conditions like sudden acute respiratory syndrome (SARS) which has a relatively high fatality rate. That was a vivid reminder that **airborne infectious diseases can pose a worldwide pandemic threat.**

Microorganisms are present in the air, not only during times of publicized infectious disease breakouts, but continuously as a result of normal people coughing and sneezing or just simply breathing. These 'bugs' are always in the air about us, in one form or another. Our immune system by and large protects us naturally, until something more virulent comes along or until our healthy defenses become inadequate for one reason or another. Therefore, we should never take the ever present risk of airborne infection for granted. It is a real concern that we cannot afford to ignore.

From these organic challenges we turn to briefly mention a purely physical contaminant that leads to harmful radiation, up close and personal. This is not from some man-made nuclear device, or even some manufactured appliance like a microwave oven. It is as old as the earth but unevenly distributed across its surface. It's a relatively rare and inert element called 'radon.' You should know something about it.

Radon, a naturally occurring gas usually derived from radioactive decay in soil, can enter your home through possible cracks in the foundation floor and walls, drains or other openings. An estimated one out of every 15 homes in the United States has radon levels above the EPA's recommended action level. The consequences of exposure to radon progeny (or by-products) can be very serious. In fact, **indoor radon exposure is estimated to be the second leading cause of lung cancer (after cigarette smoking)**. The National Research Council has estimated that radon is responsible for between 15,000 and 21,000 lung cancer deaths in the United States each year.

Speaking of smoking, **environmental tobacco smoke (ETS)**, otherwise known as "second-hand smoke", is another major indoor air pollutant. ETS is just one source of **fine particulates** that also include household dust and fibers, which together are ubiquitous in the indoor environment. Such tiny particles when inhaled, can trigger airway reactions, but some are also able to penetrate deep inside the lung and cause serious damage to lung tissue, including cancer.

Airborne infectious diseases can pose a worldwide pandemic threat.

Tobacco smoke also contains thousands of *chemicals*, including 200 known poisons such as formaldehyde and carbon monoxide, as well as more than 40 carcinogens. Anyone living in a home or working in an office or factory building, with others who still persist in smoking, is susceptible and even prone to suffer health consequences. In the extreme, ETS causes an estimated 3,000 lung cancer deaths in America each year, and 35,000 to 50,000 heart disease deaths in non-smokers, as well as 150,000 to 300,000 cases of *lower* respiratory tract infections in children under 18 months of age. In such an environment, there is hardly any stronger case for indoor air purification to be made anywhere!

Environmental tobacco smoke is just one source of an array of products which give rise to volatile organic compounds.

Volatile organic compounds (VOC's, for short) refers here to the many chemicals used in the manufacture of building materials, interior furnishings, textiles, office equipment, cleaners, personal care supplies and pesticides. They are described as 'volatile' because these chemicals evaporate into the air at normal room temperatures. Obviously, they are therefore a major concern in indoor air quality.

Researchers from the EPA and others, have found that VOC's are common in the indoor environment and in fact, at levels from ten to thousands of times higher than outdoors. In a study of 174 homes in the Avon district of England, researchers found that the total VOC concentrations of indoor air was usually about ten times higher than outside.(5) Moreover, during and for several hours immediately after certain activities, such as routine paint stripping like any of us would do, the indoor levels could be as much as 1000 times the background outdoor levels. Analysis of indoor air samples do show anywhere from fifty up to hundreds of individual VOC's, most of which are found to be emitted from indoor materials and processes. Some of them can produce odors at very low levels and that's usually an objectionable sign, but a useful tip-off for intervention and hopefully, corrective action.

These chemicals are also referred to as 'organic' to imply that they generally contain carbon. (*Wellness buffs should note that there is no immediate association with 'organic farming' or 'organic food'*). But since carbon burns, many of these compounds, including of course organic solvents, turn out to be flammable.

What may be more surprising is that VOC's can be found in small but appreciable amounts in the air we breathe-out. It is now known that normal human breath can contain a mixture of several hundreds of these unnatural chemicals. In fact, researchers published in the very respectable *Lancet* journal, some results of a cross-sectional study which led them to propose a possible, if not practical method of lung cancer detection using VOC's.(6) They pointed out that patients with lung cancer exhale more VOC's than those without that condition, and identified 22 VOC's that specifically could be used to help in its detection. That makes a strong point, even if much better methods are available for this type of cancer diagnosis.

The health consequences of exposure to VOC's have been increasingly documented in recent years. Many are now known to be irritants and can result in early symptoms like eye, nose and throat irritation, headaches, difficulty breathing, skin reactions, dizziness, fatigue and much more. **Some VOC's are known to be highly toxic and long-term effects may occur after repeated exposure.** These effects could include cancer as well as damage to the heart, liver, kidneys or the central nervous system.

By now you're getting the picture...biological pollutants, radioactive radon, ETS, VOC's! There's more stuff in the air inside your home or office than meets the eye. We have only mentioned in passing some of the work done on the unhealthy **gases** from inappropriate combustion in wood or coal stoves, gas range appliances, space heaters, furnaces and fireplaces. The gaseous pollutants from these sources include some of the more prominent culprits in the indoor air, like carbon monoxide (which can cause asphyxiation and even death in the extreme), nitrogen dioxide and sulfur dioxide (which are mainly respiratory irritants). That's just to get us started. We have also not mentioned here the invasion of **dust mites** by the millions. This army of little menacing critters surrounds us in our beds, our furniture, our carpets ... creating an unseen world of allergens that makes millions into victims of chronic allergies and more. Then there are the favorite **pets** in the home. They provide another major source of indoor air pollution, no matter what pet lovers say or do.

All this and more is the subject of the following chapter. We will discuss each class of contaminants in much more detail there. But clearly, the problem of indoor air pollution is real and relevant. It is a nightmare to all who become aware of its challenges and consequences.

However, the awareness of air pollution did not originate indoors. Furthermore, sooner or later, the pollution that makes the news outdoors could (and probably would) find its way indoors just by opening windows and doors. So, let's look at that for a minute.

THE STATE OF OUTDOOR AIR

Ever since the 1960's, outdoor air pollution has become a central environmental issue with broad impact right across American society and indeed the world. Some countries are more notorious than others. Politicians and journalists, manufacturers and industrialists, health professionals and every aware consumer have all had to address this major development, in industrialized countries at least. Recent public awareness and a call for action, new 'clean air' legislation, pressure from the media 'watchdog', plus industrial restructuring and technical innovation, have all contributed to improving the air quality from those terrible days only a generation ago. Much progress has been made.

Indeed, we may have forgotten what the air was like back in the 1960's. Pollution streamed from American factories and cars, while heavy clouds of smog settled helplessly over many urban centers. Do you remember when 'acid rain' was the talk of the town? It was not unusual to find evidence of 'soot' on one's clothing at the end of the day. To blow into a tissue coming home would often reveal the signs of black particle pollution that our airways had contended with throughout the day. Just imagine that more than 150 people died in New York City, following a 1966 Thanksgiving "killer fog" as a result of just breathing that city's noxious air.

By the end of that decade however, the federal government had introduced Clean Air legislation (1967) and the Environmental Protection Agency

(EPA) was created (1968) by presidential order. President Richard Nixon warned the nation shortly after the first Earth Day in 1970, that without far-reaching reforms the United States would, by the year 2000, become "a country in which we can't drink the water, where we can't breathe the air, and in which our children ... will not be able to [enjoy] the beautiful open spaces ...of the American landscape." We could not blame the messenger that time. The message was terrible but true. Something had to be done, and it was.

We began to clean up factories, automobiles and leaded gasoline. Industrial polluters could no longer freely dump toxic clouds into the air or hazardous effluents into the waterways. They now had to make sure that they met new national standards based on the public health impact. Car manufacturers developed catalytic converters and cleaner cars that could run without billowing uncontrolled noxious fumes across the nations roads and highways. This made the skies above the nation much more friendly and not daily darkened with smog and soot. Widespread problems like airborne lead and carbon dust were grossly diminished.

But those developments were not the complete answer, not then and not now.

Much more remains to be done. Old coal-fired power plants, for example, continue to pollute the air leading to serious health consequences. New analysis has estimated that such power plant pollution contributed up to 24,000 deaths in 2004, and some 550,000 asthma attacks, 38,000 heart attacks and 120,000 hospital admissions.

Although further action continues to generate major controversy between the considerations of politics and economics, private and public interests, and the short-term and long-term consequences, we must continue to do more. We cannot afford to be negligent - whether or not the nation becomes a signatory to the International Kyoto Agreement or adheres to the follow-up Montreal Accord which both seek to address these problems on a worldwide scale.

Identifying pollution: 'Soot' and 'Smog'

The American Lung Association (ALA) released its *State of the Air 2005 Report* which gave "good" and "bad" grades to the nation. It negligently focused only on the air outdoors, the traditional concerns of the general public. And even then, the reasons for concern remain. Clearly, outdoor air pollution levels have improved recently but it continues to take a tremendous toll on the nation's health. There are many millions of people, of all ages and backgrounds, who still live in areas where pollution in the air makes it difficult and sometimes even dangerous to breathe.

The ALA Report focuses on what they consider to be the two most pervasive air pollutants: high levels of ozone (a major component/indicator of 'smog') and particle pollution (otherwise experienced as 'soot'). These are clearly not the only air pollutants outside, but they are perceived to be the two most dangerous ones because of their presumed toxicity and their prevalence. Of the two, particle pollution is clearly the more deadly.

Millions of people in the U.S. live in counties where environment monitors show unhealthy levels of outdoor air pollution in the form of either short-term or year-round levels of particle pollution or excess ozone. The data indicates that 152 million Americans - 52% of the population - live in areas where they are exposed to either of these, whereas 52 million - nearly 17% of the population - live in areas with unhealthy levels of all three.(1)

'Soot'

Particle pollution is still a widespread problem, particularly in large parts of the eastern states, California and Texas. The term generally refers to any combination of fine solids and aerosols that are suspended in the air we breathe, inside or outside. Sometimes you may not even be aware when you're inhaling such particle pollution. But make no mistake about it, that could be really hazardous to your health. It could even take years off your life.

A key defining characteristic of airborne particles is simply their physical dimensions or size. Some airborne particles are so fine, they can only be seen with an electron microscope. Yet some are large enough that you see them floating in the air when a beam of sunlight comes streaming through a narrow opening or window. You certainly see them as the occasional haze formed by millions of particles blurring the spread of sunlight in the atmosphere.

Size is very important, especially when you inhale particles. The largest ones don't get past the body's natural defenses. You cough or sneeze them out of the body or they just get trapped in the nostrils or throat until you choose to dispose of them. But anything less than about 10 microns in diameter (equivalent to about one-seventh of the diameter of a typical human hair) gets past those airway defenses and ends up trapped in the lungs. The very smallest particles go further and manage to cross over from the alveolar air pockets in the lung and end up right in the bloodstream just like oxygen molecules do.

But particles in the air are not exactly what they seem. Some are tiny solids just like you would imagine, but others are truly liquid or else solids suspended in liquids (aerosols). It all depends on where they came from and how they are formed. The larger, more solid particles are formed *mechanically* as materials are broken down into smaller and smaller bits until they are dispersed like a fine cloud into the air. Examples of this may arise from incomplete combustion of coal and oil (including components of automobile exhaust), dust storms, construction and demolition, agriculture and household dust.

Most of the fine and ultra fine particles in the air, on the other hand, are formed as a result of essentially *chemical* processes in the atmosphere. During normal combustion, gases are produced and in the mixture, oxygen and water vapor combine with carbon, nitrogen, sulfur and volatile organics to produce a host of other compounds. These coalesce and condense to become a wide variety of new compounds as particle mixtures and aerosols. Examples of these derive every day from the burning of fossil fuels in factories, power plants, steel mills, smelters, diesel- and gasoline- powered vehicles, plus the burning of wood in residential fireplaces and the burning of agricultural fields and uncontrolled forest fires.

Whatever the sources of this particle pollution in the atmosphere outdoors (or in the air inside), there is no remaining doubt about the disastrous impact on human health. During the last two decades, there have been dozens of short-term community health studies done in many US cities and indeed around the world, to prove unequivocally the detrimental health effects associated with increased particle pollution. These health consequences range from increased respiratory symptoms, to increased emergency room visits and hospitalizations, to increased mortality from both respiratory and cardiovascular disease.

Those were the short-term results. The classic *Harvard Six Cities Study* reported in 1993 documented the major risks to human health from lifelong exposure.(7) In six small towns across the eastern United States, researchers demonstrated that there was increased risk of premature death from particle pollution in the most polluted areas, compared to the least. This was later confirmed by others from the large nationwide database of personal histories from the American Cancer Society. **All told, tens of thousands of Americans die prematurely due to exposure to particulates each year.** The pathologists report that chronic lung damage from particle pollution is not unlike that found in the autopsied lungs of cigarette smokers.

The most definitive epidemiological evidence that long-term exposure to air pollution is associated with life-threatening diseases, including lung cancer, was provided by a team of researchers from Canada, New York and Utah.(8) They followed about 500,000 adults for 16 years in 156 cities and found that people living in hazy cities across America were more likely to die of lung cancer, heart attacks and respiratory failure than people in communities with cleaner air. This effect is comparable in scope to the hazard posed by exposure to secondhand smoke (ETS).

More and more studies implicate 'microscopic particles that form haze' as a serious health risk. But for decades now, cleanup has been focused on ways to reduce ozone, as though ozone was somehow the responsible culprit in mediating the health consequences of air pollution. It is still popular to think 'ozone' when you think of 'smog' and air pollution. But it is still incorrect to do so. There is much more to 'smog' than you would be led to believe from the media reports.

Anyone living in an area with a high level of particle pollution is obviously at risk. The same is true of any type of air pollution, including what you see and what you cannot see. Let's look a bit more closely at the notorious 'smog'.

'Smog'

Much of the air pollution that is visible, is essentially particle pollution. But you need to know 'the rest of the story' for it is just as deadly. Often, *it's not only what you can see that threatens you, but also what you cannot see.* What you do not see in the dirty clouds of billowing smoke coming from factories, smoke stacks and chimneys all across the nation's industrial heartland, and what you do not see in the rolling black clouds from forest fires or the sputtering exhaust trails from cars and trucks on the highway, is the invisible mixture of volatile gases and organic compounds that are also produced from the burning of fossil fuels like gasoline or coal and more particularly indoors, from the evaporation of fossil-fuel-based chemicals like paints, cleaners and many consumer products.

On any given day when the 'smog level' is high and everyone it seems is sounding the alarm, the air is pervaded with all types of hydrocarbons, oxides of carbon (including both carbon dioxide and monoxide), oxides of nitrogen and sulfur, as well as a range of other volatile organic compounds, some of which are more prevalent and hazardous than others.

That list is conspicuously incomplete and for good reason. There is one critical "pollutant" that was not mentioned - and that was intentionally so. Here we must be very careful. **From the media and other interest groups, there has come an unreasonable and dangerous preoccupation with a single ingredient of this ubiquitous 'smog'.** Even meteorologists through their daily weather forecasts, and some environmentalists, media reporters and writers, give all the other components an undeserving "pass", but reserve all their alarm and condemnation for the one ingredient that has become the index of this type of air pollution. That is 'ozone'. Sometimes the term is even presumed to be synonymous with 'smog'. But it really isn't.

Let's be more explicit.

'Smog' is a chemical mixture of gases that typically forms a brownish yellow haze in the air, primarily over urban areas. T**he term is a short form of *smoke* plus *fog*.** Components of smog include not just ozone, but also nitrogen oxides, volatile organic compounds, sulfur dioxide, acidic aerosols and gases, and particulate matter. It is the product of chemical reactions between oxygen, certain airborne pollutants and strong sunlight, particularly the ultraviolet component. It is most prevalent during the summer months when there is typically more sunlight and the temperatures are the highest. Naturally, it increases around areas with elevated industrial and automobile emissions, and if concentrations become high enough, it can pose a significant threat to animal, plant and human life.

Ozone itself is an extremely reactive gas molecule comprised of three oxygen atoms. In contrast, normal oxygen gas molecules are very stable and have only two oxygen atoms. Ozone is formed as a product of chemical reactions when nitrogen oxides and hydrocarbons or volatile organic compounds, combine with oxygen in the presence of both heat and sunlight. In the process, an additional atom of oxygen is transferred to an otherwise stable oxygen molecule to form the unstable ozone adduct. The basic ingredients are derived from the 'pollution sources.' That is, the nitrogen oxides are emitted from power plants, motor vehicles and other sources of high-heat combustion. Similarly, the VOC's are emitted again from motor vehicles, chemical plants, refineries, factories, gas stations, lawn mowers and other such sources.

If we could remove the man-made components of smog - especially the VOC's and industrial particulates - then we would be left essentially with a natural balance of oxygen, ozone, nitrogen and nitrogen dioxide: all normalized to safe and healthy levels. But when a large amount of man-made hydrocarbons and other volatiles are added to the equation, the amounts of nitrogen dioxide and ozone are artificially increased. It's a man-made problem of pollutants for which ozone formation is, in part, a natural attempt at solution. Why? Because ozone itself reacts with pollutants in the air in an attempt to neutralize them. **Ozone is therefore, in one sense, like a thermometer (reflecting), and in another sense, like a thermostat (regulating) the undesirable air pollution that we generate.**

We will come back to this subject in much more detail later.

THE GREATER '*INDOOR*' DANGER

Air pollution does not cease to be a problem as you enter your home, office or factory floor. In fact, in real terms, the problem becomes intensified. Although most public and media attention has been focused on outdoor air pollution, the fact remains (and it is worthy of reiteration) that we spend 90% of our time inside. To make matters worse, we now know that there is often more contamination of indoor air, despite all the popular images of industrial smoke stacks or automobile exhaust and congestion.

In the 1980's, the U.S. EPA's *Total Exposure Assessment Methodology* (TEAM) *Study* provided a model for comprehensive assessment of the contributions of both indoor and outdoor exposures to total personal exposure.[9] And what did it show? This study yielded the conclusion, which was surprising at the time, that ***indoor pollution sources*** **were generally a far more significant contributor to total personal exposure for toxic volatile organic compounds,** than are emissions released by some industrial sources into outdoor air.

Another landmark investigation of outdoor air pollution that we referred to earlier, *the Harvard Six-Cities Study,* also proved to be invaluable in understanding residential indoor air pollution and its contribution to total personal exposures for a number of pollutants, including particles and nitrogen oxides, among others.[7] In this community-based study, the researchers used the combination of outdoor and micro-environmental monitoring and personal exposure assessment to characterize the contributions of various indoor sources to total personal exposure.

The sources of indoor pollution are different for rural and urban areas. In the country, as in the developing world, the main pollutant sources are from human activities such as cooking, especially when using bio-mass fuels such as wood, kerosene and dung. That's in addition to off-gassing from potent building materials and of course, the ubiquitous smoking, including 'second-hand' smoke.

A report from the World Health Organization addressing the links between indoor air pollution, household energy and human health in developing countries, put the focus on the pollution produced indoors by burning biomass fuels and coal in simple stoves with inadequate ventilation. It was estimated that **indoor air pollution is responsible for nearly two million deaths annually in those developing countries** and around 4% of the burden of disease.[10] There is much evidence that this kind of exposure increases the risk of acute lower respiratory infections in children, chronic obstructive lung disease in adults, and lung cancer where coal is used extensively.

But indoor air pollution is not just a third world problem. It is a real global concern. Back in 1992, **the World Development Report concluded that indoor air pollution is 'one of the four most critical global environmental problems.'** [11]

In the large urban centers of America - like all the other sprawling metropoles of the industrialized world - buildings are known to have major pollutants including nitrogen dioxide, carbon monoxide, radon (from building materials, water and soil), formaldehyde (mainly from insulation), asbestos fibers, mercury, man-made mineral fibers, volatile organic compounds (VOC's), allergens and tobacco smoke - as well as health damaging microorganisms like bacteria, viruses, mold and mildew. Most of the health consequences arise from materials that give off these pollutants, especially smoke, radon, asbestos, formaldehyde and other VOC's.

But in either case, one cannot neglect the invasion of the indoors by all the known pollutants on the outside. As we observed before, opening and closing doors and windows and the flow of human and material traffic dictate that what's *outside* will make its way *inside*. 'Smog' is not just an urban cloud or dense traffic signature. The handwriting is also on the interior walls. **There is no safe haven when it comes to air pollution.** We are hounded by airborne contaminants everywhere. Sooner or later we must stand up and face the enemy.

Composition of Clean Air

This book is about indoor air pollution, but we must consider what ideally constitutes normal clean, fresh air. We've almost forgotten that species.

Normal air is a mixture of gases: mainly nitrogen (78.08%), oxygen (20.95%), argon (0.93%) and carbon dioxide (0.03%) plus a number of other gases present in trace amounts (totaling less than 30 parts per million). We must also include water vapor which is present in varied quantities and typically defined by the extent of saturation or atmospheric humidity.

There is no safe haven when it comes to air pollution.

We might just take that for granted but in cosmic terms, that is a very peculiar and unique atmosphere for a planet of this size. Pure chemistry would dictate that for a typical small planet, only the most common and unreactive medium-sized molecules should remain. These might include nitrogen, carbon dioxide, ammonia and methane (which is commonly found in the atmospheres of nearly all other small bodies in the solar system). Lighter gases like hydrogen and helium escape the weak gravity but are more common on giant planets and form the bulk of the Jupiter atmosphere, for example. Highly reactive molecules like oxygen or fluorine combine with rocks and are quickly removed from the atmosphere. But on the earth, in particular, the presence of liquid water dissolves carbon dioxide and ammonia; living organisms remove most of these two gases and methane from our atmosphere, and photosynthetic plants keep a huge surplus of oxygen in the air, replenishing it as quickly as it is removed by chemical combination with rocks or more importantly, by respiration of living organisms including *homo sapiens*.

This latter respiration process combines with the photosynthesis of plants to cycle oxygen and carbon dioxide in a pathway of Intelligent Design. Outside air, while it is often polluted to the point of being unhealthy, has a distinct advantage in that nature will rebalance it as much as it's at all possible. Plants do that and as we shall see later, ozone and ionization processes in the atmosphere do exactly that. But indoors … that's another matter. There is no "nature" as such to rebalance the composition of the air we breathe there. And that's not an original idea.

Can Plants help?

Scientists at NASA and the Associated Landscape Contractors of America did a two-year study that confirmed that plants, just by themselves, help to clean the air both inside and out.(12) As a matter of fact, plants can clean up to 87% of some indoor contaminants, given the right circumstances. NASA was interested in alternatives to cleaning space station air on long duration expeditions in space. They wanted to know whether plants could be put into the orbiting space station by design, to perform multiple roles of air cleaning, oxygen production and a small amount of food production. We will learn later how serendipity led them to new space-age innovation that is the basis of the best air purification technology known to be available anywhere on earth today.

The results were very impressive and the message for the civilian public is that we could help to clean the air inside our tight buildings on earth, with special plants that can save energy, money and the pollution of the environment, while keeping ourselves healthier. The plants would not only absorb chemicals from the air, but also help balance the humidity and increase oxygen levels that stimulate healthy respiration.

'O for a breath of fresh air!'

ARE YOU AT RISK?

In some ways, we are all prone to denial, especially when it comes to matters of illness and health. We seem to be somehow predisposed to think that anything might happen or anything could be wrong with 'anybody else but me,' at anytime. Men in particular, are very good at this, but in a bad way. They tend to be slow to visit the doctor, slow to get reading glasses, slow to concede chest pains, slow to take their vitamins, for example. When it comes to air pollution, women perhaps are just as guilty. They want to make sure that their homes and places of employment are clean and safe. Who wants to acknowledge that the surrounding air leaves much to be desired? Does that not reflect on the person(s) responsible? Just look around, one might say, and see how clean and tidy this place looks.

But is that all there is? Most likely not. Appearances can be deceptive. What looks 'clean and safe' might by no means be. There are obvious exceptional hazards and risks of exposure to intense toxins that are regulated by the OSHA (Occupational Safety and Health Administration), the EPA and other environmental agencies. Unless there is some particular accident or dangerous threat that would usually involve local communities as a whole, the vast majority of us are unlikely to suffer from such hazardous exposures. But that's not the concern here. What you are more likely to do battle with, are the insidious, commonplace offenders which degrade the daily quality of life and over years of constant exposure can affect your state of health and in the extreme, even threaten to kill.

This is the present challenge in indoor air quality. We are surrounded by many such offenders in our homes and workplaces which threaten us in significant ways, as silently as the very air we breathe. Fortunately, we can do something about all of these but we must first become aware of our inherent risks so we might then seek to take control and do something to improve the otherwise dangerous situation.

To give you a better idea of what your real risk from indoor air pollution might be, here is a short and simple questionnaire, just to get you thinking and to guide your personal assessment.

A Personal IAQ Questionnair

1. **Are the buildings you spend most of your time in, relatively air tight?** **Yes** ☐ **No** ☐
   Since the 1970s, buildings have been built or remodeled to be tight and conserve energy, but they pose a substantial increased risk of having polluted air. If buildings have frequent drafts, they might be energy inefficient and derive more pollution from the exterior environment. But they obviously allow lots of air changes and are not as prone to indoor air problems.

2. **Do those same buildings provide inadequate ventilation?** **Yes** ☐ **No** ☐
   You will know from the experience of stale air and sometimes uncomfortable breathing. Some types of equipment may have additional venting needs such as dry cleaners, printers, garages, laboratories, bakeries, salons etc. and should comply with local codes and guidelines from the OSHA (Occupational Safety and Health Administration).

3. **Is there frequent use of solvents, sprays, paints, glues or strong cleaners?** **Yes** ☐ **No** ☐
   There is a wide variety of these toxic substances that form a regular part of your indoor environment, especially in the workplace. You cannot take these chemicals for granted.

4. **Do you have any associated environmental-type symptoms?** **Yes** ☐ **No** ☐
   These might include burning eyes, headaches, runny nose, itchy irritated skin, malaise or any kind of breathing difficulty or disturbance associated with any building location. You may consider the effects of 'stress' at work or even at home since indoor air might not be the only culprit.

5. **Is there sometimes an unusual or distinct odor?** **Yes** ☐ **No** ☐
   This could be more obvious in newly built or renovated buildings which tend to have more polluted air after the workers have finished their job. Many building materials and furnishings give off chemical odors which represent harmful degassing.

6. **Is there ever a smell of burning or smoke?** **Yes** ☐ **No** ☐
   This could derive from poorly ventilated furnaces, wood stoves, gas appliances and any type of heating that is not properly maintained. Incomplete combustion indoors is a major environmental hazard. Regular servicing is a must.

7. **Are you in an area where radon poses a significant threat?** **Yes** ☐ **No** ☐
   It's always wise to have your buildings tested for radon if you have any lingering doubt. Naturally, some areas and local communities may be at increased exposure and must be monitored aggressively. Ask around in your neighborhood.

8. Are there any signs of mildew or mold anywhere? Yes ☐ No ☐
Increased humidity can cause moisture condensation especially on windows and walls, which favors the growth of microbes, with the characteristic odors of mold and mildew. Similarly, if carpets or furnishings get wet, that also poses a threat.

9. Is your neighborhood known to have polluted outside air? Yes ☐ No ☐
This is especially relevant in urban areas and around industrial plants and institutions which burn-off their wastes - smoke-burning stacks, congested traffic areas and any form of high density living with fewer trees and limited open space - all lead to increased pollution outside.

10. Does anybody smoke in your building? Yes ☐ No ☐
Both the direct effects of smoking as well as the exposure to second hand smoke are known to be very harmful to health. The Surgeon General has spoken eloquently for decades now and little more needs to be said. If you or yours still smokes, it is more than time to quit.

11. Do you use volatile consumer products? Yes ☐ No ☐
If you use paper towels, wax paper, paper grocery bags, facial tissues, permanent press clothes, carpeted floors, pressed wood or particle board furniture, natural gas … you've got it.

12. Do you live or work near a parking area or heavy automobile traffic? Yes ☐ No ☐
Automobile emissions have come a long way since the days of leaded gasoline and low efficiency engines. But the most efficient cars today are still polluting the air we breathe. If you have inside access to an attached garage, you're not immune either.

13. Do you have a fireplace, wood-burning stove or natural gas appliance? Yes ☐ No ☐
You are exposed intermittently to potential particulates and carbon monoxide pollutants from incomplete combustion. Even with careful operation and experience, the risk is never eliminated.

14. Have you recently moved, renovated or refurbished your home? Yes ☐ No ☐
New homes, carpets, furnishings continue to give off dangerous vapors and fumes that might reassure some people that they're moving up. But they do much more than that. Cleaning agents, solvents, new wallpaper and the like all add to the look, the odor and the pollution in the air.

15. Do you have any pets? Yes ☐ No ☐
Pets are notorious for contributing allergens, odors and other contaminants to indoor air. Their dander, hair, urine, feces - are all elements to contend with.

16. **Do you have any unusual indoor hobbies?** **Yes** ☐ **No** ☐
Activities like jewelry making, woodworking, pottery, oil painting, etc. can be sources of airborne particles, volatile organics and more.

17. **Is your furnace rigorously maintained?** **Yes** ☐ **No** ☐
Furnaces should be inspected and cleaned yearly by a licensed heating contractor. It involves more than just replacing the filter.

18. **Is dust noticeable on your furniture?** **Yes** ☐ **No** ☐
The dust that collects in your home is just a symptom. It is indicative of the airborne particulates that float in the air consistently. Even for the most meticulous housekeeper, dust is unavoidable.

19. **Do you have rugs or broadloom in your home?** **Yes** ☐ **No** ☐
Carpeting is a sink for airborne pollutants. Dust mites, allergens, aerosols, odors and much more accumulate there. Regular vacuuming is mandatory but always incomplete.

20. **Do you use professional lawn care services?** **Yes** ☐ **No** ☐
The use of commercial pesticides outside puts all house occupants at risk. Human and pet traffic, as well as wind currents, provide these chemicals easy access to the indoors, if they are not rigorously controlled during use.

If you answered YES! to any of those questions, you might be anxious to read what follows in the remainder of this book. You will learn about the indoor air contaminants that you cannot avoid and the health consequences that they can cause which you cannot afford. That's all described here in Part One, which only addresses the Problem.

Later in Part Two, you will learn of **FIVE ESSENTIAL STRATEGIES** that together comprise a practical and effective Solution. In particular, you will learn how **FIVE CRITICAL FEATURES** come together in the latest space-age technology for air purification.

Finally, in Part Three you will learn how this technology is applied to purify the air in the **FIVE IMPORTANT SPHERES of Activity** that you engage in indoors. You'll be all the wiser and healthier for it! So keep going.

"There's so much pollution

in the air now

that if it

weren't for our lungs,

there'd be no place

to put it all."

Robert Orben quotes
(US magician and comedy writer, b.1927)

THREE

Ten Contaminants You Can't Avoid

"It's not just what you can see that threatens you, it's also ***what you cannot see****."*

We have already noted that indoor air pollution raises serious concerns that you cannot afford to ignore. Whether it be the U.S. Environmental Protection Agency, the American Lung Association, the World Health Organization or any of the many reputable institutions that emphasize this serious public health challenge (if not crisis) today, the message is abundantly clear. Environmental pollution must be among our highest priorities, both personal and national, private and public. Air pollution is clearly a central consideration in this area and since we spend more than 90% of our time indoors where polluted air is more the norm than the exception, the focus of our attention (and hopefully, research too) must be directed there.

Rember, the fact is that it's not just what you can see that threatens you, it's also what you cannot see. "With outdoor pollution, you can see the sources," says Dr. Philip Landigran of Mt. Sinai School of Medicine in New York, an expert on environment and health. "By contrast, you are looking at millions of individual sources (affecting indoor air). It's a regulatory nightmare."(1)

In the previous chapter, we identified general classes of indoor pollutants which included biological contaminants, volatile organic compounds and fine particulates of household dust, fibers and cigarette smoke. In addition, we mentioned radioactive radon, combustion gases, airborne dust mites and ubiquitous pets. That's by no means, an exhaustive list. Each indoor environment will have its own particular challenges for the occupants to face.

In this chapter, we want to take a closer look at the individual classes of pollutants that determine the poor quality of indoor air which now pervades many of the buildings - both residential and commercial - in the United States today. These are contaminants you cannot avoid. The ten leading contenders (in no preferential order) are listed below:

Ten Major Indoor Contaminants

1. **Environmental Tobacco Smoke**
2. **Volatile Organic Chemicals**
3. **Pesticides**
4. **Pollen**
5. **Dust mites**
6. **Combustion Products**
7. **Radon**
8. **Pets - dander plus**
9. **Dust fibers**
10. **Microorganisms**

Any of these contaminants would provide enough reason for concern. Individually, they each pose a problem for modern urban dwellers particularly. Trapped in our energy-efficient buildings and our fossil-fuel-loving automobiles, we fall victim to the adverse health effects of the inevitable contamination in the air that any of these can cause. They will be present to different degrees, in different locations and at different times, but you will for sure encounter one or more.

So, let's find out a bit more about these enemies we all must face. We will begin with perhaps the most obvious contaminant of indoor air, one that is clearly preventable if we would only choose responsible behavior.

EVEN SECOND-HAND SMOKE IS DEADLY

The Surgeon General has convincingly warned us that active smoking is the major preventable cause of morbidity and mortality in the United States today. It still accounts for about 400,000 deaths each year.

However, **50 million Americans continue to make smoking their personal lifestyle choice**. After all, we live in a free society. But freedom is not absolute. Smoking does have secondary effects on those who do not smoke but must sometimes share the same polluted air.

Since the 1960s as we have become more increasingly aware of air pollution, we have found that involuntary exposure to tobacco smoke is also a significant cause of preventable morbidity and mortality in non-smokers. There is voluminous evidence of such exposure and its consequences.[2,3]

Non-smokers inhale second-hand what is called Environmental Tobacco Smoke (ETS). This is the combination of the *mainstream* smoke exhaled by the active smoker (how **disgusting!**) and the *sidestream* smoke that is released from the burning of the cigarette (how **disturbing.**)

Tobacco smoke is a complex mixture of particles and gases that contains hundreds of chemical species. Smoking indoors increases the levels of particles that can penetrate the lungs: addictive nicotine, toxic polycyclic hydrocarbons, carbon monoxide, acrolein, nitrogen dioxide, and many other harmful substances. The net increase clearly varies with the number of smokers in each place, the extent of smoking, the ventilation of the area and the use of air cleaning/purifying devices. Exposure in the home is worst for young infants who do *not* attend daycare, whereas for adults residing with non-smokers, the workplace may be the principal location for ETS exposure. A U.S. nationwide study in 1993/94 showed that average exposures in the home were generally much higher than those in the workplace.(4)

Second-hand smoke is "a major preventable cause of cardio-vascular disease and death."
- American Heart Association

Transportation environments, especially private cars and trucks, may also be heavily polluted by cigarette smoking, but fortunately, smoking is now entirely banned in all domestic North American flights and many international ones too. However, in the privacy of our homes, we can still make our own lifestyle choices. It is estimated that environmental tobacco smoke still contaminates almost 50% of all U.S. homes and a significant fraction of our workplaces.

In both active and involuntary smokers, the smoke components and their metabolites are detectable in body fluids (blood, urine, saliva) and in alveolar air (in the lung). Such data provide evidence that involuntary exposure to ETS leads to absorption, circulation and excretion of tobacco smoke components. They confirm a high prevalence of involuntary smoking. Studies in adult non-smokers exposed to ETS at home, in the workplace and other settings, show levels of contamination ranging from 1-8% compared to the related active smokers.(5) In children, their levels reflect the behavior of their smoking parents or caregivers.

Unfortunately, some of the worst consequences of indoor air pollution with ETS are observed in innocent children. They fall victim to the unsociable and addictive habits of parents and others in the same household. Active smoking by mothers in particular, leads to a variety of adverse health effects in children. Pregnant women who do not smoke but are exposed to ETS can also have those pollutants cross the placenta with fetal consequences.

Children are not the only ones to suffer from ETS. Adults do too. In 1981, researchers in Japan and Greece reported evidence to indicate increased lung cancer risk in non-smoking women who were married to cigarette smokers.(6) Since that time, other investigations in the United States and elsewhere have added to that evidence. The association of involuntary smoking with lung cancer is still somewhat controversial but it seems very plausible, given the presence of carcinogens in sidestream smoke and no obvious threshold dose for the carcinogens that the active smokers breathe in. Passive smoking must therefore remain stigmatized and considered an important cause of lung cancer in public health terms.

In a similar fashion to lung disease, coronary heart disease is also associated with exposure to environmental tobacco smoke. More than 20 studies of the association between ETS and cardiovascular disease have been published.(6) The American Heart Association's Council on Cardiopulmonary and Critical Care has concluded that ETS increases the risk of heart disease and "is a major preventable cause of cardiovascular disease and death."(7)

What's the real picture? Involuntary exposure to tobacco smoke, as in the polluted air at home or office, threatens the fetus in the womb, causes respiratory infections in children, increases the prevalence of respiratory symptoms, reduces the rate of functional growth as the lung matures, and causes lung cancer, heart disease and more in non-smokers. **The estimate of the number of adult deaths in the United States from second-hand smoke exposure is about 46,000 per year,** including 3,000 from lung cancer, 11,000 from other cancers and 32,000 from heart disease.(6)

A word to the wise should always be sufficient.

FLOATING CARCINOGENS EVERYWHERE

Environmental tobacco smoke is but one source of carcinogenic pollutants in the indoor air. Hundreds of volatile organic compounds (VOC's) have been identified in environmental tobacco smoke. As we mentioned in the previous chapter, the typical air in most buildings is infiltrated with any range of these chemicals that are derived from many sources, both outside and inside. As a class, VOC's are the most prevalent indoor air pollutants. They are also the most studied.

VOC's are in widespread use as industrial and domestic solvents, in paints and coatings, in pressed-wood products, fragrances and a host of other practical ingredients in processes and consumer products. They are big business. Entire industries like petrochemicals and plastics are based on these chemicals and they are not going away - at least, not in the foreseeable future. They represent one of the burdens of industrialization. According to the U.S. EPA, of the 3000 chemicals produced in the United States today at over 1 million pounds per year, almost 500 are identified as consumer product chemicals. We cannot avoid them.

Indoor air samples may contain twice the number of VOC's and at levels several times higher than outdoors. We get exposed to them daily by constant inhalation, with serious possible consequences. In the 1980s the U.S. EPA carried out the TEAM studies as mentioned in Chapter 2 to measure personal exposure. They selected about 750 people, representing 750,000 residents in areas scattered across America. Each participant carried a personal air monitor which had a battery-operated pump that pulled some 20 litres of air across the sorbent matrix over a 12 hour period. The major findings may be summarized as follows(8):

1. Personal exposure exceeded median outdoor air concentrations by factors of 2 to 5 for nearly all the prevalent VOCs, even in areas that were highly industrialized or had heavy vehicular traffic.

2. The major sources identified were consumer products (e.g. bathroom deodorizers, moth repellents); personal activities (e.g. smoking, driving), and building materials (paints, adhesives).

3. Traditional sources (automobiles, industry, petrochemical plants) contributed only 20-25 percent of total exposure to most of the target VOCs.

A major category of human exposure to toxic or carcinogenic VOCs that you are likely to encounter is that of room air fresheners and bathroom deodorants. Beware! Since the function of products like this is to be effective for extended periods, they are designed to maintain elevated concentrations in the indoor air at home or in the office. Therefore, extended exposure to these associated VOC's is most likely to be the highest pollutants among non-smoking persons.

Detecting VOC's in the human body provides a direct indication of the total dose of the pollutants through all the important environmental pathways. For many of them, the most sensitive method of detection is by analyzing the exhaled breath but this tends to measure only the most recent exposure. Blood measurements have detected VOC's at occupational levels for many years. This has been done for a dozen or more of the most prevalent VOC's, including the carcinogenic solvents, benzene and chloroform.

We now know that **indoor sources account for somewhere between 80 and 100% of the total airborne risk associated with these chemicals.** The notable exception is carbon tetrachloride which has been banned from consumer products entirely by the U.S. Consumer Product Safety Commission.

Some VOC's like benzene and vinyl chloride, have been proven to cause cancer in humans. Others have been proven to cause cancer in animals but are only presumed to cause cancer in humans. Still others are mutagens or just weak animal carcinogens. Although the risk estimates may remain somewhat uncertain, prudence would dictate that we try to avoid all VOC pollutants.

IN FEAR OF PESTICIDES

Pesticides represent another class of important indoor air pollutants. A pesticide can be defined as '*any substance used to control, repel or kill a pest such as an insect, weed fungus, rodent etc.*' Conventional pesticides are mainly semi-volatile or nonvolatile organic compounds. They are quite dangerous and as such the U.S. EPA regulates their sale, distribution, use and eventual disposal.

Conventional pesticides are used both indoors and outdoors in homes, office buildings, schools, hospitals, nursing homes and many other public facilities. Anyone can find a variety of pesticides to choose from in the local hardware or department store. Various preparations are readily available 'off the shelf' for looking after common flies, cockroaches, ants, spiders and moths within the home. You can find sprays and shampoos for dealing with fleas and ticks in pets; insecticides for use on houseplants and home gardens. Avid gardeners will want herbicides, insecticides and fungicides for their gardens and lawn treatment. Most homemakers themselves apply disinfectants routinely to clean bathrooms and kitchens, deodorize rooms and spray laundry. Professional pest control and of course lawn maintenance are also big business.

Studies conclude that about 90% of all U.S. households use pesticides in one form or another. The major exposure of the general population to these chemicals occurs inside the home. Results from simultaneous 24 hour indoor air and personal exposure monitoring, showed that 85% of the total daily adult exposure to airborne pesticides was from breathing air inside the home.[9] Just to emphasize the magnitude of this, consider the sales figures. According to the U.S. EPA, direct purchases of conventional pesticides for home and garden use accounted for 17% of the $12 billion spent on pesticide products in the US in 1997. Non-agricultural commercial sales made up 13% and most of that was intended for home, office and institutional application.

Even at the best of times, you will get exposed to airborne contamination by conventional pesticides. However, the misuse of pesticides by people in the home as well as by commercial applicators happens much too frequently. Every day, hundreds of alarm calls and emergency visits take place, reporting suspicious pesticide poisoning. But that's only the tip of the iceberg. What is of much greater concern, should be the chronic effects on health associated with long-term exposure in indoor environments where these chemicals are trapped in human lungs. It is more than enough to know that **a number of *pesticides* have been classified by the U.S. EPA as known (Group A), probable (Group B) or possible (Group C) human carcinogens.** That is indeed enough to cause fear. Users, beware!

Conventional pesticides may be described as semi-volatile or non-volatile, but when they are applied indoors they do tend to become airborne to a significant degree. Most are semi- volatile. They may be periodically introduced into the indoor air by direct application as insect sprays, disinfectant sprays or room deodorizers. Some vaporize from treated surfaces like carpets and baseboards and get re-suspended into the air in particles. They can accumulate in upholstery and in or on children's toys. Others applied to the foundation or perimeter of a building as in common practice, can leak or track into the interior spaces. Sometimes when used outdoors, pesticides can be blown inside or taken in on clothing and shoes to accumulate as house dust and be re-suspended in air. Pesticides may even come in on manufactured products like carpets, furniture and paints.

Disinfectants like pine oil and phenols (including BCP and PCP) represent the most common household pesticides. They are quite common environmental pollutants and constitute a significant component of indoor air. Acid herbicides are among the most heavily used household pesticides. Although they are not frequently reported in indoor air, they are commonly detected in house dust.

A number of pesticides are known carcinogens ... Users, Beware!

The concentrations of pesticides in indoor air will vary significantly in time and space. They will be highest just after and near points of application. Concentrations typically drop rapidly for about three days after application as the pesticide is absorbed into furnishings or dissipates to the outdoors with ventilation. However, even at 21 days later, concentrations of the more volatile ones may still be as much as 20 to 30% of what they were on the day of application.[10] Most household applications are done during the warmer spring and summer months when pest infestations are greatest. The warmer weather may increase volatization to further pollute the indoor air.

Speaking of warmer weather, the thoughts of spring remind us of allergy season and the distress of allergy sufferers who must contend with airborne pollen. That's our next airborne contaminant.

POLLEN IN LOVE AND WAR

Pollen is a fascinating study. With its *sex* and *violence*, it should make a popular prime-time documentary for TV. Let's just tell the story here.

Pollen is indispensable to the life cycle of flowering plants. Plant life (and perhaps human life too) would be almost impossible without its movement from one plant to another. The pollen grain is a highly sophisticated capsule that protects the male gametophyte which gives rise to "sperm" as it is transported from the anther where it was produced. It arrives at the receptive stigma in style and grandeur, to be welcomed by a guard of honor that forms a 'pollen tube' leading to fertilization of the awaiting egg by the sperm. That, in a nutshell, is how flowering plants do their thing. Forgive the analogy, but it's a kind of *flowering sex* ... plant style.

Let's be more explicit!

Some plants have their choice pollen carried by other living species (active transport) but others just leave their pollen grains to be carried by air currents (passive transport) on their romantic journey. They are literally blown in the wind. These latter flowers, unlike their other cousins, don't have to attract any other species. So they tend to have smaller, less busy petals and sepals and little attractive 'juices'. Those would only be impediments to windborne transport of pollen. Instead, they specialize in larger anther filaments extending up into the windstream and knowing their reduced chances of making a successful 'hit', they produce many more pollen grains per plant just to compensate. These grains tend to be smaller but more hardy. Nature loves to play around but it does not fool around. You get the picture. **It is these airborne pollen grains that are most likely to find their way into human nasal mucosa and cause allergic symptoms.** In such cases, the romance ends not in ecstasy or comedy, but in tragedy. The allergy sufferer is distressed.

This romantic endeavor is quite affected by climate conditions, especially temperature and moisture. Most plants in temperate regions release their pollen over only a few weeks each year, when it is warm and dry. They are true 'spring lovers', coinciding with the April showers and the May flowers. In warmer climates, the flowering season tends to be longer but you will still see seasonal cycles for pollination which remain predictable. The beginning of the pollen season is usually on or about the same day each year. Ragweed loves to do its thing in the daylight, and requires about 14.5 hours of daylight for the flowering to start. It sheds during the early to midmorning hours but disperses the pollen only with sufficient air movement (wind). Rainfall effectively 'scrubs' or 'works out' any pollen from the air, thereby raining on its grand romantic parade of destiny.

Air currents blow pollen as they will, sometimes right through open doors and windows or else through cracks or gaps in the outer walls of buildings, as well as through air intakes. Pollen can also enter buildings with people and pets. At times, this could surpass the wind blown pollen. Carriage on clothing has been demonstrated particularly on farms. Of course, one cannot exclude indoor sources from houseplants, especially in greenhouses. Whatever the mechanism, indoor pollen can be as much as half of what it is outdoors. It tends to be higher if windows and doors are left open, and much lower when air conditioning removes bioaerosols from room air. Under certain conditions, dust-borne pollen and its allergens could potentially contribute to allergy symptoms.

When pollen reaches its choice destination to fertilize other flowers - that could be considered a form of 'making love'. But when it gets to humans inadvertently, it 'makes war' with the human immune system. That's the *violence* referred to above. Upon contact with the mucosal lining of the nasal passages, the eyes or even the oral cavity, pollen grains and their allergens encounter mast cells which declare the war. In sensitive allergy sufferers, they trigger the release of histamine and other substances which produces the typical annoying and sometimes debilitating allergy symptoms.

Pollen is just one of the allergens, which in turn is just one of the triggers that exacerbate asthmatic conditions. Such hypersensitivity typically leads to wheezing and difficulty breathing.

There is no doubt that pollen is a major factor in the typical hay fever and allergic asthma epidemic seen in the Spring each year. Most of it is probably generated outdoors and clearly, exposure to pollen indoors must be reckoned with. A large portion of the population suffers from hypersensitivity to the wide variety of airborne pollen types. For them, pollen is a wicked, unrelenting enemy. Ask any allergy sufferer who can hardly sleep at night, especially during the Spring. At such times they must think that 'the war will never end'.

DUST MITES BY THE MILLIONS!

Allergy sufferers anticipate the increased distress produced by pollen in the Spring and then another major bout in early Fall. But in the home, there are challenges all year round. A wide variety of foreign proteins float around in the air and in the common dust in homes. These are often potent antigens that induce immune responses in sensitive inhabitants. Being indoors is not real relief for the allergy sufferer.

It has been known for a long time that household dust is a major cause of allergic symptoms and disease. Studies have shown that allergic sensitization to the house dust mite is the most common cause of allergy in asthmatics in most parts of the world. Some 60-80% of such patients are affected. Until quite recently, household dust extracts have been made from the contents of vacuum cleaner bags and widely used for routine skin testing and immunotherapy. More recently, many of the relevant proteins themselves found in house dust have been purified, cloned and sequenced in the laboratory. Antibodies have been produced which can be used to quantify these proteins both in the air and on dust samples.

The protein allergens found in the home come principally from domestic pets, dust mites of the genus *Dermatophagoides*, and the German cockroach. Cat and dog allergens become airborne on small particles which tend to fall slowly, stick to walls and clothing, and so become widely distributed in the community. On the other hand, dust mite antigens are found mainly on larger particles which stay airborne only briefly in a local environment and are not spread widely. They accumulate in carpets, bedding, furniture and the like.

Many, many years ago, dust mites were found principally in birds' nests, but now they have found a better place to live. They have made their way into our warm, snug and humid houses. In fact, dust mites were first recognized in bedding back in the 1920's but it took another 40 years to establish their role as a source of allergens and a trigger of asthma. Skin testing has now established this all over the world. The specific mite allergen has been identified as a protein labeled as *Der p1* which can be quantified to establish a universal standard.

But dust mites themselves are **living creatures**. They are microscopic, eight-legged scavengers that live in the common household dust, mainly feeding off the tiny flakes of human skin that we shed all the time, and constantly breed in our carpets, bed linens, curtains and furniture. These dust mites belong to the *arachnids* family, which also include spiders, chiggers and ticks. They are really tiny and it takes 200-250 of them in a row to make one inch, or 1000 to fill a teaspoon. They cannot be seen with the naked eye, but they prove to be fascinating under the microscope. They have oval bodies covered with ultra fine striations. Perhaps what's most remarkable is the observation of their little legs which have small sticky pads to enable them to burrow deep into carpet fibers and fabrics so that they can resist the pull of powerful vacuum cleaners. In other words, they not only hang around, they insist on staying there!

And they do have a rightful place. They are a natural and perhaps necessary part of the food chain. As scavengers, they go about their daily task of cleaning up our environment at least on their terms. The presence of house dust mites should not be equated with a lack of cleanliness or poor hygiene. Each mite typically lives about 2 to 4 months and produces about 20 droppings per day. It is these droppings which contain the protein allergens (*Der p1*) that prove to be such a nuisance and distress to allergy sufferers.

If just the thought of these little monsters is making you squirm, then considering the numbers will make you paranoid or even drive you crazy. It is estimated that an ounce of dust can be host to 10-20,000 of these creatures. Then consider that in a single year, the typical home can generate up to 30-40 pounds of dust. Do the math and you will readily appreciate that **the typical house is occupied by dust mites in the millions, defecating everywhere! Yikes!**

Imagine your average bed with 10,000 of these creatures making up *their* beds. Your only reassurance is that they do not bite, sting or transmit disease, so they do not really pose any threat except to people who are allergic to their fecal protein. That includes about 10% of the population.

You're probably anxious by now and wondering what you can do to fight back. That comes much later in the book but just to allay your fears for now, here are your TEN TIPS for controlling dust mites:

TEN TIPS For Controlling Dust Mites

- Encase mattresses and pillows with dust proof or allergen impenetrable covers.
- Use a dehumidifier or air conditioner in the summer to maintain lower relative humidity. Dust mites thrive in moist air.
- Wash all bedding/blankets in hot water once a week. This will kill the buggers!
- Wash all blinds and curtains regularly.
- Replace wool and feather-stuffed bedding materials with synthetics.
- Keep only the washable stuffed animals and toys.
- Choose hard flooring in preference to carpets, especially in bedrooms.
- Use a damp mop or rag for dusting. Dry cloths only stir up the mite allergens.
- Vacuum only with good ventilation (during and after). Central vacuums which vent to the outdoors are preferred. Portable vacuum cleaners must have highly rated filters.
- Use an efficient air filtration/purification system.

DANGEROUS COMBUSTION

We mentioned in Chapter 2 the major worldwide problem associated with the burning of biomass fuels and coal in simple stoves with inadequate ventilation. **In the Developing World, this accounts for some 2,000,000 deaths each year.** [(11)]" But even in industrialized countries like ours, heating and cooking still demand the burning of fuel, which produces smoke and fumes. This gives rise to emissions with irritating properties and which can prove very toxic if ventilation is inadequate. In temperate climates, the challenge is to keep heat in, while getting fumes out. That sometimes proves to be more difficult than one would expect. Central heating avoids some of the problems of smoke posed by open fires beneath chimneys and in wood-burning stoves. But the latter are still common today.

Combustion is itself a complex chemical process which gives rise to an array of products. The most obvious is visible smoke which is laden with fine particles and has an odor that many find romantic in winter. As would be expected from elementary chemistry, burning hydrocarbon fuels (which obviously contains hydrogen and carbon) would lead to combination with oxygen to yield the corresponding water (H_20) vapor and carbon dioxide (CO_2). These would be relatively harmless except that water vapor contributes to dampness in a house, which has implications for mold growth and dust mites that can lead to health consequences. Carbon dioxide is relatively harmless except perhaps in the event of accidental fires when it could help cause asphyxiation.

If hydrocarbon fuels were pure and the process of combustion were complete, that would be the end of the story. But again, it just isn't so. Incomplete combustion leads to other products like carbon monoxide, oxides of nitrogen and sulfur and some volatile organic compounds that themselves could prove deadly.

Carbon Monoxide (CO)

Carbon monoxide is a colorless, odorless gas formed by incomplete combustion of carbon fuels. It is exceedingly toxic and remains a major concern of both indoor and outdoor air pollution. Here again, the focus has been on the outside and many countries have set pollution standards for vehicle emissions and industrial plants, while the indoor problem has been neglected. But not by everyone.

It is not too uncommon (for men especially) to park the car in a closed garage with the engine running, only to be discovered later as another suicide victim. Fatal carbon monoxide poisoning is much lower in women and is mostly accidental. Usually, accidental CO poisoning is associated with fires. But malfunctioning of combustion sources and operations with inadequate ventilation remain the most common causes of acute CO-poisoning. Most times, it is non-fatal. In the home, it is usually some kind of heater gone bad. Victims might be using piped gas, coal, butane or heating oil, when something goes wrong.

In an American study, 6 out of 14 homes containing **kerosene heaters** had CO-concentrations that exceeded the maximum 8 hour levels specified by ambient air quality standards which limits CO to a level of 9 ppm (parts per million). Examination of blood samples from the residents showed increased CO-hemoglobin when the heaters were in use compared to what was found when they were not.(12) Could this be typical of many homes across America? Probably so.

That's not without health implications. The big problem with carbon monoxide is that it gets into the lungs, then into the blood where it attaches to hemoglobin tenaciously, taking the place of essential oxygen. In fact, it has an affinity 250 times that of oxygen, so it is not displaceable. It therefore limits oxygen delivery to the tissues and even impedes the release of the oxygen that does get there. On exposure, the first symptoms of CO-poisoning are usually headache, fatigue and impaired exercise tolerance. With prolonged exposure or higher concentrations, visual changes, dizziness, nausea, vomiting and diarrhea set in. Then later there would be confusion, collapse, coma, convulsions, respiratory failure and death. One unfortunate scenario is seen when a person is near collapse, yet is unaware of anything wrong. Finally, with a little more exertion, collapse is sudden and immediate, with no chance of escape. Even when CO-poisoning is not so severe, there can be long-term chronic neurological and psychiatric changes, at least.

Nitrogen oxides (NO_2)

Nitrogen is the major constituent of normal air. During combustion, with all that's happening chemically in the heat and flames, nitrogen and oxygen can react to form a range of oxides, some of which are also detectable in normal air.

In the home, the major sources of nitrogen oxides (principally, nitrogen dioxide) are gas-fueled cookers, fires, water heaters and space heaters. Pollution occurs if the emissions are not removed to the outside, either because there is no venting system or what is available is inadequate or not working well. We also cannot neglect the NO_2 that can enter the home from an attached garage, but it is fortunate that unlike CO, the oxides of nitrogen are produced during acceleration and much less while the car is just idling. There is also a small contribution to NO_2 from cigarette smoking.

Fortunately, low levels of NO_2 as found in the U.S. (compared to significantly higher levels in Europe), cause only minor degrees of ill health. These are mainly respiratory symptoms, susceptibility to respiratory infections and some limited reduction of lung function. Infants and children appear somewhat more susceptible.

Sulfur dioxide

Sulfur is a trace component of many fuels, including coal, kerosene heating oil and other fossil fuels. When these are burnt, the sulfur combines with oxygen in the air and a significant amount of sulfur dioxide (SO_2) is produced. This sulfur dioxide is a well known pollutant of the air outdoors from industrial coal and oil combustion and because of its known environmental damage, it is carefully regulated and monitored as a pollutant.

The concentration of SO_2 in indoor air is generally much lower than that outdoors, partly because it is well absorbed onto room surfaces. A small number of homes in the Netherlands were identified with higher indoor concentrations relative to outside, as a result of a faulty heater or malfunctioning chimney. So that can happen. A couple of American studies found some excessive levels in homes with kerosene heaters. Those seem to be a particular class of culprits.

Most of what we know about the health effects of SO_2 comes from studies in the external air. We know that SO_2 is very soluble in water and tends to irritate the moist mucous membranes of the eyes, nose and throat. Sometimes, it can be carried by water droplets or carbon particles right into the terminal bronchioles of the lungs, to aggravate respiratory disease. Chronic effects, as in industrial exposures, lead to decline in lung function.

Other combustion products

The burning of wood, coal and oil gives rise to other products besides these gases just described. Some of the particles and volatile organic compounds produced, including **polycyclic aromatic hydrocarbons (PAH)**, can also have significant health consequences, including some demonstrable cases of cancer in animals.[13] Little is known of safety limits of all these combustion products in the indoor environment, but since people typically spend so much time indoors and have such long-term exposure, it is more than prudent to minimize any potential risks to health. Those already in poor health or those more vulnerable to respiratory diseases ought to be particularly concerned. More research is needed in this area.

THE RADON CONTROVERSY

Radon is a noble and inert gas which has been identified in indoor air for half a century. It is a proven carcinogenic source which has led the U.S. EPA to stipulate control guidelines for builders and to call for voluntary testing of essentially all single-family residences in America, so that some mitigating action can be taken if annual average concentrations exceed a specified threshold. That is a costly proposition so the policy is not without its antagonists and the debate continues to this day.

But radon has an interesting history. In the late nineteenth century, miners working in the Erz Mountains of eastern Europe showed unusually high rates of cancer. High levels of radon were first identified in the mines in that region which made it at least a suspect in the case. By the 1950's, the epidemiologic studies showed convincing evidence that radon was indeed guilty.

Well, not exactly. Radon is formed when naturally occurring, radioactive uranium decays. This ^{238}U is universally present in the earth, so radon is being formed everywhere. In mines, it enters the air from the ore itself, or it is brought into the mine dissolved in water. But radon itself is not stable. It emits α- particles (helium nuclei) with a half-life of 3-5 days. In other words, every 3-5 days *half* of whatever radon is present is converted to other species (progeny) which, in this case, turn out to be solid. The progeny (which can be thought of as daughter nuclei or offspring elements) then form small molecular clusters or attach to aerosols in the air. When these are inhaled (now hold your breath!) they are deposited on the epithelial lining of the lung and then they themselves emit more α– particles. These particles have limited penetration into tissues, but enough to damage the cellular genetic material, giving rise to cancer. At least 20 different epidemiologic studies of underground miners confirm that radon exposure increases the risk of lung cancer.(14)

In the case of private homes, the radon enters as soil gas by penetrating through cracks and crevices, or around concrete slabs. It may also sometimes get in by building materials or through water, especially from deep drinking wells. Radon levels of concern have now been identified in homes in North America as well as Europe. More recently, research has focused on the need to better understand the risks posed to the general population by indoor radon. All the risk assessments to-date, based on case-control studies, have shown that radon progeny should be considered a significant cause of lung cancer in the United States.(14) The value of that risk is consistent with what the data from studies of miners would suggest.

Monitors have been devised for relatively inexpensive and convenient measurement of indoor radon levels. Passive units utilize alpha-track detectors, activated carbon monitors or electrets which monitor accumulated charge.

The truth is that radon concentration in homes varies widely across the country - from one region to another and from one house to another in the same region. This reflects both local characteristics of the soil and different housing characteristics. Naturally, homes with basements have the largest potential for higher concentrations. The bottom line is that there are as yet no accurate techniques to predict any category of homes which are more likely to have high radon concentration. It's no wonder that the U.S. EPA has called for voluntary testing everywhere.

Radon causes one-third of all lung cancer deaths.

If that sounds like an over-reaction, just consider that **one-third of all lung cancer deaths are attributed to radon concentration levels that exceed the current U.S. EPA guidelines.**(14) But of the radon attributed deaths, only a minority can be prevented by current risk management strategies since we still have to win the unseen battle against indoor radon pollution.

Control programs focus at present on identifying homes with levels above the stipulated guidelines and then taking remedial action to lower them. To prevent the entry of radon from the soil, some techniques are now being applied: soil depressurization, sealing, building pressurization and source removal. On the other hand, to remove radon after it has already found its way indoors, the two principal responses are: expanding ventilation and utilizing technologies for cleaning the air.

Still the issues of radon and lung cancer remain controversial, simply because of the challenge of indoor radon. As a society, we've yet to engage the real enemy: Indoor Pollution. Radon is just one of the issues, but a real one.

PETS SOME LOVE TO HATE

No matter how you explain it, Americans love their pets. It is estimated that more than 70% of U.S. households have one or more domestic pets, most likely a dog or a cat. Other fairly common pets would include birds, hamsters, rabbits, mice, gerbils, guinea pigs and even rats. The pet population well exceeds 100 million. Even in the impersonal Internet Age, pet lovers still form strong attachments to their pets, especially the children, singles and seniors who enjoy the comfort, security and companionship that pets afford. Children often learn a sense of responsibility and lessons about life and death from their pets. Seniors, on the other hand, appreciate their pets which help alleviate the so-called 'empty nest syndrome'

But that's only one side of the story. Some 10 million pet owners are known to be allergic to their treasured animals. There is no more compelling sign of love for their favorite pets than the sacrifice they make every day just to enjoy having them around. A study from Japan recently revealed that one out of every four patients with pet allergies continues to keep a pet, despite their obvious allergies and asthma-related problems. What devotion! Or is it better to say, what folly?

For many years, pet-loving patients and their physicians have known that pets are a major source of allergy symptoms. Their dander, or skin flakes, as well as their saliva and urine can cause significant allergic reactions in about 10% of the general population. Although the animal hair itself may not be directly implicated, the pet's hair does collect pollen, dust, mold and other allergens.

The major allergens associated with cats and dogs have been identified, characterized and most thoroughly studied. A special nomenclature identifies these as Fel d 1 (for *felis domesticus*, the cat genus) and similarly, Can f 1 (for *canis familiaris*, in the case of dogs). These known proteins may come directly through the skin (probably from the sebaceous glands), and appear in the animal's body fluids (saliva, urine). They have been shown to be present in the air of most pet inhabited houses.

Moreover the pet allergies become attached to fine particles (less than 5 microns) which tend to remain airborne for extensive periods. The particles tend to be sticky, so they are found in large quantities on clothing, walls, furniture, carpets and bedding. This becomes a key method of transfer throughout a community. Several findings have confirmed that **pet allergens are readily taken from house to house, with or without pets.**

Many allergic individuals report onset of classic symptoms (runny nose, irritated eyes, wheezing or simple malaise or headache) within 15-30 minutes of entering a house where pets live. The characteristic odors in such homes also confirm airborne chemical sensitivities that many find uncomfortable, irritating and not infrequently, obnoxious.

So pets pose their own challenges to those individuals sensitive to their airborne allergens. As such, they present a significant source of indoor air pollution. But the devotion of pet lovers must constrain any strategy to control this problem.

More of that later.

FIBERS IN RANDOM FLIGHT

Many an indoor environment is contaminated by dust. Of course there is the dust generated from human skin flakes, water droplets and aerosols, or fine particulates from environmental tobacco smoke and combustion of fuels. Then there is the drifting dust blown inside from the dispersion of dirt and sand related to vehicular traffic, wind gusts and construction activity. But, in addition to all that, many products made from fibrous substances are distributed everywhere in present day buildings.

These materials are used mainly for thermal insulation and acoustic (noise) applications. Examples of insulating products would include blown-in wall insulation, blanket insulation in attics, pipe and exterior duct insulation and boiler jackets. Acoustic products include ceiling tiles, wall panels, interior duct lining and treatments applied in the ceiling to absorb noise. Today all these products are made from vitreous fibers and cellulose (wood) fibers, but up to 1980, it was common to use asbestos. Large surface areas of these fibrous materials are found inside as part of the air distribution system of buildings. There they are exposed to ever present air currents and there is always the risk that these fibers will be disseminated throughout the building and cause health consequences to inhabitants. This is of particular importance today to building maintenance workers, and we already know of the price paid by earlier asbestos workers.

Asbestos

Asbestos refers to a class of fibrous silicate mineral materials that were once popular for their proven resistance to heat, moisture and chemical agents. The small size of many of these fibers allows them to penetrate deep into the alveolar region of the lung. Chronic irritation of the lung tissue leads to scarring and fibrosis (asbestosis) with obvious decline in lung function. Asbestosis is an occupational disease seen in asbestos workers but also in household (family) members of these workers. Building maintenance workers whose duties involve work on or near friable asbestos-containing materials are also at increased risk.

The incidence of lung cancer is increased in workers chronically exposed to asbestos, and this is clearly made much worse for those who smoke cigarettes. There is a particular cancer of the lining of the lungs (the pleura) called *mesothelioma* which is almost exclusively associated with this kind of asbestos exposure. These kinds of risks, however, are finite (four cases in a million) but very much reduced for incidental, low-level exposure as in the case of occupants in buildings with asbestos, or just living with asbestos workers. Such risks are well below the levels associated with radon exposure in community buildings. However, public concern and activism pushed the US Congress to enact comprehensive legislation requiring identification of all practical materials which contain asbestos, and an active in-place management plan in all school buildings. (*Where's the corresponding radon legislation? Is it only 'the squeaky wheel that gets the grease'*?)

Vitreous fibers

Since about 1980, asbestos has been replaced as the fibrous component of insulation and acoustic materials by synthetic vitreous fibers and cellulose. The vitreous fibers are formed by cooling molten liquid mixtures containing oxides of silicon, aluminum and phosphorous (derived from sand, rocks and slags). These fibers are widespread today, found in many common building materials, like spray-applied fireproofing, ceiling tiles, thermal insulation, sound insulation, fabrics, filtration components, plasters and treatments for acoustic surfaces. The possible erosion of fibers from the parent materials into the air-stream of buildings is of significance to indoor air pollution.

Since the mid 1980's, vitreous fibers have been suspected as a possible cause of some symptoms associated with 'sick building syndromes'. These symptoms were primarily skin irritation and the irritation of mucous membranes. They have been implicated also as a possible cause of outbreaks of 'office eye syndrome.' But the fibers affecting the skin tend to be somewhat larger than 4 micron diameter only. It seems that that irritation is due to local mechanical or physical effects as the fibers are transferred from surface to skin and skin to mucous membranes. However, an airborne transfer route from surfaces to mucous membranes does result from disturbance of the settled fibers as a result of activity of building occupants. The point is, that whereas asbestos fibers are smaller and remain suspended in the air to travel large distances via the air distribution system, vitreous fibers are generally larger and usually settle out more readily.

The bigger threat is posed directly to workers in facilities where vitreous fibers are manufactured and handled in construction. Such workers are more likely to inhale these fibers. The continuous filament glass fibers have not been classified as a human carcinogen. Glass, rock, slag wool and ceramic fibers are classified as 'confirmed animal carcinogens with unknown relevance to humans.' Several epidemiologic studies have shown statistically higher risks of lung cancer and other respiratory system cancers among workers employed in fiber manufacturing.[15] The reviews of airborne vitreous fiber levels measured in buildings concur that public health risks for cancers from projected lifetime fiber exposures are below reasonably estimable levels.

Cellulose Fibers

Since asbestos is a known human carcinogen and vitreous fibers have been classified as possible carcinogens, the so-called 'healthier alternative' has become cellulose. As such, recycled newsprint has been used ingeniously to make thermal insulation for buildings. Some even favor it as a 'greener alternative' since it uses much less energy and reuses 'precious trees' rather than add to carbon fuel combustion.

Again the adverse health effects are seen in cases of occupational exposure rather than in occupied buildings per se. The cellulose can become a nuisance dust, and the fibers may cause irritation to the mucous membranes and the upper respiratory tract. However, cellulose is used as a building material in many applications, like in ceiling tiles or thermal insulation. It tends to retain moisture, which favors microbiological growth. One type of mold that causes upper respiratory tract irritation and pulmonary hemorrhage in children, has been shown to grow readily on wet cellulose material.

Cellulose fibers are derived indoors not only from building materials, but also from other commonly used paper products. It is therefore not surprising that concentration of cellulose fibers in indoor air far exceeds that of asbestos and vitreous fibers. Wood dust (saw dust) is composed mainly of cellulose and other carbohydrate fibers. The International Agency for Research on Cancer (IARC) has classified wood dust as a known human carcinogen (carpenters, beware!).

BEWARE! MICROBES IN THE AIR

We need not become excessively paranoid, but whenever we stop to think about it, we recognize that we are surrounded by a world of microorganisms that share our space and environment. The air we breathe, the water we drink, the food we eat, everything we see and touch is populated with these small life forms that are struggling for their own survival, and that they do, often at our expense.

Bacteria, viruses, fungi are our constant companions. Most of the time we can live together in health and harmony. Our immune systems are developed to protect our own interest so that we can remain healthy and thrive in this natural ecological balance. But then, when certain pathogenic invaders threaten us, we could be overwhelmed and succumb to disease.

You may call them 'microorganisms', 'microbes', or just plain 'germs', but they have their rightful place in the world. They make the soil fertile; they change, and often improve our food; they clean up the environment, and perhaps most importantly, the best of them protect us from the worst of them. We speak of 'friendly bacteria' in our digestive system, for example, and every time someone dear to you responds dramatically to penicillin, you should give thanks for the fungus mold that made it possible.

But they can be dangerous, and sometimes deadly!

Many microorganisms can survive airborne transmission and in a more confined local environment indoors, the risk of the spread of disease is multiplied. Therefore, no consideration of indoor air pollution can neglect the major problem of microbial contamination. These miniscule particles of living matter can be driven or carried from place to place, indoors or outdoors. They can go along by themselves as clusters in many cases, or they can hitch a ride on other floating particles or aerosols just the same. Clearly, given any index case of infection, normal breathing, speaking, coughing or sneezing will propel bursts of microorganisms into the air, which drift across any distance indoors to be inhaled by unsuspecting occupants.

In hospitals, this is of paramount importance and healthcare professionals spare no initiative or expense to practice infection control by sophisticated, proactive designs and response protocols. Despite their best efforts, however, patients in hospital are at increased risk of acquiring what are called 'nosocomial' infections. These are infections acquired while in hospital or other healthcare facilities resulting from exposure to organisms that are transmitted in one way or another. Airborne transmission is certainly a possibility, just as by other means like food and water, surface contamination, oral-fecal, IV lines and catheters, diagnostic procedures, surgery … the list goes on. Certainly, modern hospitals make this prevention a top priority. Yet it happens every day.

But in the home or office, it is open season. One person is infected and they share their bounty with everyone around. Much of the time, this may happen *before* that person has advanced symptoms to trigger withdrawal or isolation.

The recent outbreak of Severe Acute Respiratory Syndrome (SARS) caused worldwide alarm. The virus killed 299 of the 1,755 people known to have contracted it in Hong Kong, the epicenter of the epidemic. Across the globe, 8000 people were infected and some 800 died. The disease was spread by invisible aerosol droplets floating through the air.

The outbreak brought unforgettable pictures on nightly television everywhere. Who could forget the sight of large numbers of ordinary citizens in major cities, going about their daily duties wearing protective masks to avoid contamination. It was scary, just thinking of the possibility of a deadly virus coming soon to a space near you. But then, when the excitement was all over, the media hype completed its usual cycle and we all returned to business as usual.

However, the threat of airborne disease contamination remains. From the spread of the common cold, to much more serious diseases, inhaling microorganisms carried on airborne particles and aerosols can be dangerous to your health. Many infectious diseases are known to be transmitted in this way. Some examples are given in **Table 1**.

Table 1. Examples of Airborne Diseases

Viral	Bacterial
Chicken Pox (*Varicella*)	Whooping cough (*Bordetella pertussis*)
Flu (*Influenza*)	Meningitis (*Neisseria*)
Measles (*Rubeola*)	Diphtheria (*Corynebacterium diphtheriae*)
Smallpox (*Variola*)	Pneumonia (*Mycoplasma, Streptococcus*)
S.A.R.S. (*Coronavirus*)	Tubercolosis (*Mycobacterium*)

So then, indoor air pollutants include both living and inanimate materials. Among the living, **bacteria** isolated from indoor environments, both from HVAC internals and other surfaces, as well as from air samples, have included a wide range of species. In some cases, species known to cause infections have been found for sure, but the vast majority identified are generally regarded as being harmless to humans. They range in size from about 0.5 - 5.0 microns and occur in almost every environment, but particularly in dusty, dirty places inhabited by humans or other animals. Individually, they cannot be seen with the naked eye but they can be artificially cultured by the millions to form visible colonies. Some can be the 'worst of enemies' to human health and yet others 'the best of friends'. But they are in the air.

In the case of **fungi**, the vast majority of types are of little concern except those that are known to cause allergic reactions and also infections, in persons who prove to be susceptible. Together with yeasts and molds, they are single or multicellular filamentous organisms with well defined cell walls and nuclei. They reproduce sexually and asexually, often with an abundance of spores or seeds in the air stream. They tend to enter buildings from the outdoors and set up colonies wherever they find favorable conditions for growth. Many different species are known but five generally predominate in the indoor air: *Stachybotrys, Cladosporium, Alternaria, Penicillum and Aspergillus*. Up to 85% of patients allergic to molds will react to one or more of these.

Viruses are somewhat peculiar because they are neither living nor non-living, and yet perhaps both. Yes, they are hard to pin down categorically. They are so tiny, in the range of 0.02 - 0.3 microns. While floating in the air or just sitting on some surface, they are (for all intents and purposes) considered inert. They have little or no life *of their own*. But (and here's the key!) if and when they make contact with a suitable plant, animal or bacterial cell, they appear to come to life. They infect the cell, hijack its constituents, and reset the cell's agenda to facilitate viral replication. Air becomes an effective transfer medium. Certain viruses are transmitted from person to person by droplets in the air (eg., influenza, measles) or in association with materials shed from the body (eg., seeds from the pocks produced by chicken pox and smallpox viruses). Some viral species may even survive long enough in the circulating air, protected by other particles or cells, and are then able to infect other people in the same building.

Humans are themselves a prominent source of animate pollution just as we have already pointed out for inanimate contaminants. Simple movement causing your clothes to rub against your body will cause the outer, dead layers of your skin to be shed. So will just rubbing your body against walls or furniture or rubbing your hands or scratching body parts. Someone estimated that on average, we each shed about 7 million particles and cells per hour and that *each* of these might carry with it an average of 4 microbial cells. That's 28 million cells per hour or some 700 million per day. Plus, we are constantly shedding hair and losing water by breathing, speaking and perspiring.

All these processes release parts of ourselves into the indoor environment and you can be sure that every one of them is accompanied by microbes of one type or another, all derived from our bodies. Humans are veritable reservoirs of microbial contaminants of the indoor air. What we release, combined with other sources inside and from outdoors, all find conditions to set up their little colonies where 'multiplication' is always the name of the game. They can get back into the air stream and away they go, to further multiply. They love being trapped in HVAC systems and anywhere else that can support their agenda for survival.

Humans are veritable reservoirs of microbial contaminants of the indoor air.

Once these microbes encounter susceptible humans within the confines of a building, they cause trouble. They can penetrate the body's defenses and actively colonize the tissues or organs they find. This could be in the respiratory or the gastrointestinal tract, for example. The symptoms of such *infection* could be as minor as the common cold or simple eye, nose or throat irritation. Otherwise they could be extremely serious and even fatal, such as in

Legionnaire's disease where death often results from acute pneumonia.

The second major effect may be to cause *allergic reactions* in susceptible individuals. These people become sensitized to antigenic material (usually proteins) and suffer from consequent symptoms. Again, the results can be mild as in common 'hay fever' with allergic rhinitis or conjunctivitis, or as serious as seen in allergic asthma, hypersensitivity pneumonitis or 'humidifier fever."

These are just a few of the ***consequences*** of indoor air pollution that none of us can afford. These and other potential consequences are really the subject of the next chapter.

FOUR

Ten Consequences You Can't Afford

*"Many toxic effects not seen overnight, might be seen **over a lifetime**"*

Prolonged exposure to contaminants in indoor air would have consequences you can't afford. The effects may be slow and subtle, but you must recognize that over time, exposure to even very low concentrations of chemical contaminants may result in delayed toxic effects. After all, if you remain indoors most of the time and these unhealthy pollutants are trapped there, sooner or later, you will pay a price, especially if you show increased susceptibility for any reason.

SIGNS AND SYMPTOMS

Exposure to some indoor pollutants can lead to more **acute effects**. Within *minutes* there could be respiratory symptoms like coughing, chest tightness, wheezing or shortness of breath; within *hours*, one could develop all kinds of symptoms of headache, runny nose, itchy eyes and chemical sensitivity; within *days*, you might see symptoms of infection like fever/chills, a productive cough or general malaise. These are all common presentations in the family doctor's office.

Industry-related occupational hazards are generally regulated and would more than likely, be dealt with by an on-site or company physician or other health personnel. But problems arising from contaminants encountered in the daily lives of persons in their homes and offices are more likely to be seen by their primary health care provider. When the presentation and symptoms are suggestive of environmental exposure, one should assume that they more likely originate *indoors* rather than outdoors. That's where we spend most of our time and that's where the exposure is likely to be.

Unfortunately, most of what we know about illnesses and symptoms related to non-industrial indoor environments, comes by way of case studies done on buildings with persistent complaints. But we still have no clear handle on the so-called '*sick building syndrome*'. Only recently have the more sophisticated epidemiologic studies such as randomized trials been attempted. And that's understandable. These 'indoor health effects' are difficult to quantify chemically. The concentrations of contaminants tend to be small and often require complicated and expensive monitoring methods. The individual's sensitivity plays a greater role when exposed to these low concentrations compared to high concentrations. Add to that all the psychosocial characteristics and other unknown factors and the issues become increasingly difficult to study.

Unfortunately too, medical school curricula barely touches on environmental health per se. As a result, many doctors know rather little about the links between illness and indoor pollution. They too buy-in to the media myth that the problem is all outside where you see industrial and automobile pollution. What they cannot see, they are prone to ignore in this area. And that's ironic, since so much of medicine involves tracing the unseen microbiological world of activity. But the profession has been ever so slow to recognize indoor environmental illness. Instead, doctors are quick to diagnose stress, or a vague viral illness or some other infectious etiology. Fortunately, in more recent time, a number of physicians have focused on this area of 'occupational and environmental illness' and they are informing the rest of us. It is time for all the profession to catch up.

Many doctors know very little about the links between illness and indoor pollution

Traditionally, in the broadest of terms, 'sick building syndrome' (SBS) has been distinguished from the well defined '*building-related illnesses*' (BRI) that are caused by exposures to more specific indoor contaminants. Unlike SBS, BRI can be specifically diagnosed and the symptoms of BRI do not necessarily resolve when afflicted individuals leave the building (indoor) environment. Any given individual may be diagnosed with BRI without knowing the health status of other building occupants. This is not the case for SBS, where observation of the health effects on groups of occupants is essential to any assessment and diagnosis.

Building-related illnesses include airborne infectious diseases, hypersensitivity (abnormal response of the immune system) and of course, toxic reactions. On the other hand, the sick building syndrome describes a constellation of symptoms that have no clear etiology (cause) but are attributable to a particular building environment. And then there are those individuals who suffer from that other condition, now called '*multiple chemical sensitivity*' (MCS). All these are consequences of a polluted air environment, consequences you cannot afford.

But when all is said and done, pinning down a cause in the environment is often quite a challenge. Many signs and symptoms are nonspecific, making a differential diagnosis more complicated. Many pollutants cause similar manifestations and these can easily be confused with other common conditions unrelated to air contaminants. Obviously, both the acute respiratory and non-respiratory effects can be readily confused with unrelated conditions like other allergies, influenza (the 'flu') and the common cold. Many effects may also be associated, independently or in association with stress, work pressures and seasonal discomfort. In any case, whenever the environment is suspect in establishing the diagnosis or cause for patient complaints, it is important to consider the particular variables surrounding possible exposure. Some critical questions that might be entertained are listed below:

Ask The Right Questions?

- When did your symptom(s) or complaint(s) begin?
- Does the symptom(s) exist all the time, or does it come and go, associated with times of the day, days of the week or seasons of the year?
- Are you usually in a particular place at those times?
- Does the problem abate or cease, immediately/gradually, when you leave there?
- Does it recur when you return?
- Have you recently changed your place of residence?
- Have you made any recent changes in, or additions to, your home?
- Have you or anyone else recently started a new hobby or activity at home?
- Have you recently acquired a pet?
- Does anyone else in your home have a similar problem?
- How about anyone else with whom you work?
- What kind of physical/chemical/biological exposure do you have at work?
- Have you recently changed your job or job location?
- Has your workplace been redecorated or refurbished lately?
- Have you recently started working with new or differen
- Do you smoke or are you exposed to second hand smol

Those are just some of the questions you or your doctor might consider in trying to identify environmental exposure as a cause of some acute symptoms or complaints you might have. Hopefully, identifying some particular exposure as a primary cause of signs and symptoms of any acute disease condition, would lead to the removal of that source. Yet some effects can be acute, and even deadly, like Legionnaire's disease or SARS.

Over the long term, the question might become academic. You are more likely to see the effect *after* the irrevocable damage has been done. That can lead to **chronic conditions**. The obvious example is cancer. By the time the diagnosis is made, it is too late to address the possible source(s). At least for that individual patient, it is a moot point. Therefore, it is important to address the risks of exposure to environmental contaminants and indoor air pollutants, in particular, to try to anticipate and prevent the potential health consequences you can't afford.

Let's consider some of these major consequences in turn. We begin with the spread of infectious agents.

SPREADING THE COMMON COLD

The common cold is exactly that - common. A viral infection of the upper respiratory tract is probably the most common illness known the world over. It is estimated that there are one billion cases each year in America. It is the leading cause of casual visits to the doctor and of school and job absenteeism. The National Center for Health Statistics (NCHS) estimated that in 1996, there were 62 million cases requiring medical attention. There were also 45 million days of restricted activity and 22 million days lost from school. The level of morbidity for each individual may not be high, but the aggregate cost of the common cold runs into the billions of dollars - an economic cost we can ill afford.

The common cold symptoms usually manifest two or three days after infection and often include nasal discharge, obstruction of nasal breathing, swelling of sinus membranes, sneezing, sore throat, cough and headache. Fever is minimal but can climb to over 100°F in infants and young children. The symptoms typically last from 2-14 days with conservative management only (fluids, rest) but two thirds of cases recover within a week. Beyond that, other diagnoses should be considered (e.g. influenza, allergy, immuno-suppression, sinusitis, bronchitis, strep throat etc). Some of these could be complications.

At least 200 or more different viruses are known to cause the symptoms associated with the common cold. Of these, *rhinoviruses* (from the Greek *rhin* meaning "nose", therefore - nose viruses) account for approximately 30-

50% of adult colds. Rhinoviruses are known to survive up to three hours outside the nasal passages on inanimate surfaces and skin. That definitely facilitates transmission. It takes only small doses of virus (just 1-30 particles) to produce infection but yet, up to 25% of infected people show little or no symptoms. When someone has a cold, their runny nose is literally teaming with cold viruses. Sneezing, nose-blowing and nose-wiping spread the virus. You can catch a cold by inhaling the virus from the contaminated air or by touching your nose, eyes or mouth after you have touched something already contaminated by the virus.

Viruses cause infection by overcoming the body's complex defense system. That first line of defense against cold viruses is the mucus produced by the membranes in the nose and throat. It traps the materials we inhale with air: pollen, dust, bacteria *and* viruses. Cold viruses infect only a relatively small proportion of the cells lining the nose. When a virus penetrates that mucus and enters any cell, it commandeers the protein-making machinery it finds, to manufacture new viruses and the cascade begins. It appears that most of the cold symptoms we experience are probably the result of the body's immune response to the viral invasion.

The common cold is one thing, influenza is another.

THE THREAT FROM INFLUENZA

Influenza is often called 'the flu' and is caused by viruses that infect the respiratory tract. These infections often cause more serious illnesses than the common cold. Typically, the symptoms include fever, cough, sore throat, runny or stuffy nose, as well as headache, muscle aches and fatigue. The illness generally runs a one-to-two week course, but serious and potentially life threatening complications can develop, especially pneumonia in the elderly and those people with chronic health problems (heart and lung). Young and generally healthy people are much less likely to experience any complications as such and usually bounce back when 'the flu' has run its course.

In an *average* year, only about 20,000 people die across the world from influenza complications. And that's the relatively good news. The bad news is that the influenza viruses are constantly changing from year to year. The change is usually minimal so flu vaccines provide adequate (even if not perfect) coverage for those at risk from year to year. But as time goes by, new strains emerge. About every ten or twelve years, a major shift can occur with a significant change in the protein coat of a virus so that the human body experiences it as an entirely new virus. This gives rise to local epidemics.

Since 1590 there have been 10 confirmed and three possible pandemics. The most notorious was the so-called **"Spanish Flu" pandemic of 1918** which caused 500,000 deaths in the United States and more than 20 million worldwide. It devastated communities all around the world, attacking young and old alike, with shocking speed. Victims were often reported to wake up feeling perfectly healthy, and then fell ill and died before the end of the day. Others would survive the initial illness, only to develop pneumonia and die, some days or even weeks later. There were literally people dying in the streets. As the pandemic took hold, no one seemed immune and fear gripped the entire world population at the time. Then in the Spring of 1919 the epidemic faded away, just like it came, and did not return.

The 'Spanish flu' was an unusually severe influenza pandemic. It was certainly neither the first nor last one. The 'Avian Flu' epidemic of 1957-58 and the epidemics of 'Hong Kong Flu' that struck in 1968-69 and again in 1970 and 1972 demonstrate how easy it is for influenza to spread among people who lack immunity to a new strain of virus.

We have been fortunate in recent times. The influenza virus seems to adapt or mutate optimally in chickens and other birds. When they cross over to humans (and thankfully, that's rare), they pose a serious threat, especially when a new virulent strain develops. All the experts in epidemiology agree, that based on historical patterns ...

the world is now overdue for the emergence of such a virulent strain that, if person to person transmission were to occur, there would be such a devastating pandemic outbreak as to put millions upon millions of people all across the globe at risk of dying.

Today with frequent airline travel linking the global community, any such new strain of influenza could travel around the world in a heartbeat.

This is an imminent nightmare scenario for which the world remains unprepared. The threat is very real and it's a daunting challenge to develop vaccines and antiviral drugs to cope with any such eventuality. It is a sad picture, no matter how you look at it. Nobody can predict where or when the next virulent and transmissible strain of influenza will develop to start a pandemic. The warnings of a potentially imminent pandemic is based on the recent occurrence of different outbreaks of a very different **Avian (bird) flu** that jumped to humans for the first time, causing fatal infections each time.

Avian influenza is an infectious disease of birds caused by type A strains of the influenza virus. The disease, which was first identified in Italy more than 100 years ago, now occurs worldwide. Of the 15 subtypes of this virus, the H5N1 strain poses the greatest threat. It mutates rapidly, acquires other genes easily and causes severe disease. More than 70 cases of death from this threatening H5NI strain of the virus have been confirmed in Asia at the time of writing. These cases have all been reported among regular poultry workers, with their much higher risks of exposure. The sudden occurrence only since 1997 could indicate that certain flu viruses are evolving quickly enough to pose a more serious threat. The unusual Avian flu virus has shown up in chickens in several countries already, first in Asia and now in Europe (for example, in Turkey, Italy, Greece, Romania and Croatia). The first deaths from the same H5NI strain of the Avian flu virus in Europe have just been reported. Two teenagers living on a chicken farm in Turkey have recently succumbed to the virus and that only heightens the concern. The virus is on the move and with lethal force. It has now reached into Nigeria (West Africa). Some epidemiologists are already sounding the alarm.

There are international teams of scientists and healthcare professionals who are dedicated to early detection, rapid response and hopefully early elimination of such viruses. They are constantly monitoring influenza mutation and activity in human and animal populations the world over. But only one critical breakout has to occur to initiate the global threat. Fortunately, there has not been any confirmed case of **human to human transmission** of the H5NI strain to date. That would be the start of something ominous and potentially global in scope.(1)

Most types of influenza, like the common cold, are generally spread by contact, but they can also be spread directly by respiratory droplet transmission. These infectious droplets can be released into the air when an infected person coughs, sneezes, talks, laughs or just simply exhales. Just like the common cold, the droplets are too heavy to be airborne for extended periods of time and usually travel no more than about three feet before settling. If people are close enough to each other, these virus-bearing droplets can spread directly by inhalation. More often they can be transferred indirectly through touching the objects or places where they settle and then touching one's eyes, nose or mouth.

Isolation of infected persons, frequent and thorough hand washing, and wearing of protective barriers such as gloves and facial masks are classical methods (and hopefully routine ones when necessary) to try to limit the spread of infection. Later in this book, you will learn of innovative strategies to enhance such protection. Any possibility of killing an infective virus in the environment is a strategy that deserves serious consideration and careful exploration.

THE LEGIONNAIRE'S ENIGMA

The common cold may be a prevalent condition but it is generally mild and self-limiting, and therefore of little individual consequence under most normal circumstances, but for some minor inconveniences. Other airborne infections like influenza, can be devastating and even deadly. Here is a classic tale of another killer that awakened the medical world and significantly increased public awareness of the hazards posed by indoor air pollution.

The year was 1976. In the July heat that year, members of the American Legion's Pennsylvania Department and their families had gathered at the Bellevue-Stratford Hotel in Philadelphia for their 58th Annual Convention. Ironically perhaps, it was the U.S. Bicentennial Convention year for those World War II Legionnaires, so they were celebrating in grand style. About 600 of the conventioneers actually stayed at the host convention hotel. It was by all accounts, a joyous, merry-making occasion among great friends. The fellowship at the reunion was palpable and there was a parade of sorts to ignite the beaming spirits.

Just a few days into the convention, some of the more than 4,000 participants began to experience cough and fever which developed into full-blown pneumonia. No suspicions were aroused among this rather elderly crowd. The celebration carried on without pause or damper. The ill were dismissed to be cared for routinely but not carelessly. But days after the convention ended and attendees had gone home, several cases of pneumonia continued to be reported to health departments across the state. An Air Force veteran who attended the convention became the first to die at a hospital in Sayre, Penn. But that was only the first exclamation mark. A total of 221 conventioneers and other individuals who had also entered the hotel during the convention, developed the mysterious pneumonia and 34 died in total. Now there was more than curiosity. No one could give a reasonable explanation. There was a challenge to healthcare professionals to account for the local 'epidemic' that seemed to lead back unambiguously to the hotel and events there.

Even before any answers were forthcoming, the media got into the act and labeled the condition with a new name: Legionnaire's disease. After all, it was a type of pneumonia, plain and simple, except that the etiology (cause) was not yet identifiable. An extensive investigation was undertaken to determine the cause of the outbreak. Talented investigators from the Centers for Disease Control and Prevention and other researchers worked tirelessly for several months on the problem. It was not an easy epidemiological case. Everyone was confused and 'weird theories' were advanced to explain what was going on. They ranged from nickel carbon intoxication, to conspiracy theories (like "communists or pharmaceutical companies conspiring to wipe out American veter-

ans"), to specimen mix-up or willful contaminations.
alleys of speculation.

Eventually, **good science prevailed**. After months
work, Dr. Joseph McDade and the rest of his team were able t
tify some novel bacteria that were proven to be the cause of
and even fatal pneumonia.

But how did this bacteria, now called *Legionella pneumophila* come to infect the World War II veterans at their historic convention?

Dr. Carl Fliermans drew on his past experience and went poking around aquatic habitats and found the culprit. He identified the L. *pneumophila* living in natural hot water springs all over the United States and, more particularly, in air conditioning cooling towers. They had all missed an important clue. Of the 221 people who became sick around the convention, 72 were people who had nothing to do with it. But they had been inside the Bellevue Stratford Hotel or walked past it. Now, if the bacteria resided in the water of the cooling towers, it would be spread by the air-conditioning system itself, in aerosolized water droplets. People who inhaled the aerosols, actually inhaled the micro-organisms, right into the respiratory tract where they were engulfed by patrolling macrophages. The 'bugs' could remain there, safe from other hostile mechanisms of the immune system, thereby causing flu-like symptoms. If left untreated, and especially in elderly persons, they could eventually cause pneumonia and that might, not infrequently, result in death.

Eureka! Everything fell into place.

And that was just the beginning. If the same bacteria that had caused the Legionnaire's epidemic was not new, if it multiplied freely in natural fresh water sources and incidental man- made reservoirs, surely this could not be an isolated incident. It would probably have shown its true colors before but had gone unnoticed and unrecognized. So said, so done. Scientists went on a treasure hunt, digging through the medical archives in search of evidence that this bacteria had wreaked havoc before. They did not come up empty-handed. They discovered several previous outbreaks that had remained unsolved.

The most notable of these was the case of ***Pontiac fever***. Back in July 1968, nearly all 100 employees in the Oakland County Health Department in Pontiac, Michigan developed a flu like illness that resolved within a few days. But it was not influenza. No one knew at the time what it was exactly. Not until after the discovery of Legionnaire's did the CDC go back and analyze serum samples that had been frozen and saved from the infected employees. They identified *Legionella pneumophila* as the true cause of that episode too.

In more recent time, the same bacteria have caused a number of infections in rather unusual settings and the media has continued to highlight them as news. In 1989, the ultrasonic mist machine (as seen typically in the produce epartment of many large groceries and supermarkets, still often used to deceive consumers of the 'freshness' of vegetables and fruits) was shown to harbor the *Legionella* bacteria, after 33 grocery store shoppers in Louisiana were hospitalized following exposure. In 1994, it hit the cruise ship industry. There was an outbreak of Legionnaire's disease among passengers on board a cruise ship, which was later shown to originate from exposure to contaminated whirlpool baths. Then in 1996, it showed up again in a whirlpool spa display inside a popular home-improvement center in Virginia. Some 33 customers at that store developed symptoms of Legionnaire's disease and two of the victims subsequently died.

The CDC now estimates that more than 10,000 cases of this same Legionnaire's disease occur annually in the United States. The number of people who develop the milder Pontiac fever is probably larger but that is more difficult to estimate. The total number of individuals who have had exposure to *Legionella pneumophila* in the course of their lifetime is possibly in the millions. This is not an unreasonable estimate since antibody assays on blood samples of normal healthy adults show positive elevated antibody titers for the bacteria in 1-16 percent of all those tested. That is usually indicative of some prior exposure, sufficient to cause the immune system to mount a response by producing antibodies which later prove protective. At least 34 different species of L. *pneumophila* have been identified.

The course of Legionella infection has been seen to range from the mild flu-like illness of Pontiac fever to the devastating pneumonia of Legionnaire's disease. The latter extreme is obviously more frequent in older individuals (no surprise in 30 year veterans); in individuals with preexisting lung disease or poor health status; in cigarette smokers (who seem to show up as a class everywhere), and in individuals with a weakened immune system due to chemotherapy or other serious infections.

Legionnaire's disease has become the most common cause of atypical pneumonia in hospitalized patients. It is the second most common cause of community-acquired bacterial pneumonia. Outbreaks have been recognized throughout the world.

Critical elements for disease incidence usually include a mechanism for aerosolizing water from a source contaminated with *Legionella pneumophila*, and a pathway for the aerosol to reach some susceptible individuals, and then their inhalation of the aerosol.

Equipment used to manage indoor environments in buildings, and more specifically to provide cooling, has provided common artificial breeding grounds and dispersion elements. Air conditioning equipment, like cooling towers and evaporative condensers, are the type of systems that provide the warm temperatures and nutrient conditions which are ideal for the growth of the bacteria. It is then sprayed into the air. However, since 1976, air conditioning has changed. Federal agencies all over the world have imposed more stringent cleaning and hygiene provisions for cooling towers and large-scale air conditioning systems.

Low levels of the bacteria may exist in the municipal water supply. In the home, the bacteria can find niches for growth in hot-water heaters, shower heads, tap faucets, whirlpool baths and hot tubs, humidifiers and even decorative fountains. More vulnerable situations in hospitals and nursing homes require the use of sterile water in devices that supply patients directly such as respiratory equipment (nebulizers, ventilators). In many institutions, water systems are flushed and decontaminated by hyperchlorination, ultraviolet irradiation, or super heating of the water.

These issues are clearly of importance in the home, in public facilities and healthcare institutions. Although *Legionella* bacteria are ubiquitous and perhaps can never be completely eliminated from building systems (wherever there are potable-water sources, reservoirs and cooling systems), due diligence and an excellent maintenance program are mandatory personal and public health necessities, if we are to minimize these consequences we cannot afford.

Yet there are more deadly hazards from indoor air pollution.

THE MOST DEADLY OF ALL INDOOR HAZARDS

The World Health Organization points out that **tuberculosis (TB) kills more adults worldwide than does any other infectious disease (including AIDS)**. This is an airborne infection, transmitted from person to person almost exclusively indoors. It is therefore the most deadly of all indoor air hazards, at least by its consequences. These are truly consequences that no one can afford.

Most tuberculosis clearly occurs in the world's poorest areas, but it is a growing problem in many middle-income countries too. Most high-income countries, including the United States, have seen a resurgence of this disease in recent time. Today, we live in a global village with soft, permeable borders which could never contain such a highly communicable disease. Increasingly, we are seeing tuberculosis cases among people born in high-prevalence countries who have migrated to North America and Europe. At the same time, the incidence among US born residents has declined, but frequency of travel has made it easier to transport the contagion. Moreover, air travel itself has been associated with tuberculosis transmission within the confines of commercial aircraft. Infection is also possible in many other indoor spaces.

The prevalence of infection with the TB causative organism, *Mycobacteruim tuberculosis*, is estimated to include about one-third of the world's population. Fortunately, only a very small portion of this group actually have the active form of the disease. Without the active disease, tuberculosis infection is *not* contagious. Americans now experience fewer than 10 cases per 100,000 population and we contribute an estimated 5% of cases in the Americas and about 0.3% of worldwide cases. More recent infusion of resources in response to the imported resurgence of TB has done much to restore badly eroded public health programs which had been neglected for years when the problem seemed to have been under control. The major focus has been on completion of therapy for both confirmed and suspected new cases. Field trials have demonstrated that 9 months of the drug isoniazid (INH) daily (or twice weekly at a higher dose) reduces the chance of reactivation of latent infection by more than 90%, if completed.

Now, how is tuberculosis spread?

The tubercle bacilli are transmitted on aerosols which are coughed into the air by patients with active disease. The productive cough is stimulated by disease and secretions in the airways. The aerosolization process is somewhat vigorous and many bacilli do not survive their ordeal. But those that do, go for a ride across the indoor environment, only to be inhaled by any unsuspecting co-inhabitant. Such airborne organisms, if they remain viable, must get past the natural defenses of that new host's respiratory tract to land in the alveolar region of the lung. There they are likely to be engulfed by the ever watchful scavenger cells called macrophages. A contest ensues. This inherited or innate resistance allows most potential infections to be aborted early in many individuals. Otherwise, replication takes place and a more specific antigenic stimulus sets off the immune system. Acquired cellular immunity activates a more specific and lethal response. If that is inadequate, local progression takes place and that is followed by dissemination through the bloodstream. It may also lie dormant, just waiting for another day of reactivation.

Tuberculosis can affect most organs of the body. Most often it is seen as a respiratory condition with a range of symptoms including fever, fatigue, malaise, weight loss (strong clue), chronic cough (another strong clue) and less often, bloody sputum (hemoptysis). There is a common skin test for TB which you have probably done in your doctor's office. It has low predictive value in low risk populations like the United States generally. But for close contacts of an infectious case, recent migrants and travelers to endemic areas, the TB skin test usually has some useful predictive value, but only when appropriate.

In patients who show suspicious symptoms of tuberculosis, the diagnosis of pulmonary (lung) TB is confirmed by chest x-ray and more definitively, by special acid-fast smear and culture of the sputum produced by coughing. But a positive diagnosis need not strike dread in the individual today. One of the great advances of medicine in the last century was the discovery of antibiotics against serious widespread infectious diseases like TB. Treatment is possible for all index cases but the as yet insurmountable challenge of this human scourge is the widespread latency and reactivation which perpetuates the disease in high prevalence areas. In the usually associated crowded conditions, the contagious spread continues at pandemic proportions.

Tuberculosis may be a real pandemic in the developing world with all its other social, economic and health problems. We who live in western affluence, away from the widespread prevalence of such an infectious disease (although we have our own epidemics like sexually-transmitted diseases) may wish to take comfort that it is not happening here. But as global travel accelerates and migration to the industrialized West continues, we too will have to face potential consequences like TB resurgence - consequences we cannot afford to ignore.

A WAKE-UP CALL FROM SARS

Tuberculosis has been with us for generations and the sight of healthcare workers wearing gowns and breathing masks around a TB-management facility or hospital ward would be normally taken for granted. But to turn on the evening news at the dawn of the 21st century to see average citizens in modern cities in North America, Europe and Asia, commuting to work with masks covering their faces, was a reflection of the worldwide awareness and anxiety posed by the threat of a new pandemic that seemed to be just over the horizon. In the Spring of 2003, SARS had just appeared and the World Health Organization (WHO) had just declared a ***Global Alert*** for the first time in its history. "SARS" brought with it a modern 'wake-up call'.

Severe Acute Respiratory Syndrome (SARS) is a viral respiratory illness caused by a coronavirus, called *SARS-associated coronavirus* (SARS-CoV). The novel diagnosis was first reported in Asia in February 2003, after appearing in Southern China in November 2002. It was recognized soon after as a global threat in March 2003. Over the next few months, the illness spread to more than two dozen countries (suspect cases were reported in 31 countries) in Asia, Europe, North America and South America before the SARS global outbreak of 2003 was effectively contained. More than 8000 people worldwide became sick and approximately 9% of them died. Fortunately, in the United States, only eight people had confirmed laboratory evidence of SARS-CoV infection, all of whom had traveled to other parts of the world where SARS was known to have been present and transmission was occurring. There was no evidence that SARS spread any more widely in the community in the United States. No one died here.

What are the signs and symptoms of SARS?

The illness usually begins with a high fever (measured above 100.4°F or 38°C) which is usually associated with chills or other symptoms, including headache, general feeling of discomfort and body aches. Some people also experience mild respiratory symptoms at the onset. Diarrhea is seen in approximately 10-20 percent of patients. After the incubation period of some 2 to 7 days, SARS patients may develop a dry, non productive cough that might be accompanied by or progress to a condition of hypoxia (low oxygen blood levels). In 10-20 percent of cases, patients require mechanical ventilation. Most patients develop pneumonia. Intensive and supportive medical care is the primary therapy, as no specific treatment has yet been shown to consistently improve the outcome of the ill SARS patient.

Previously identified human coronaviruses (named for their spiky, crown-like appearance under the electron microscope) were known to cause only mild respiratory infections. The symptoms of SARS are similar to those of influenza, but clinicians have developed a reasonably fast and accurate test to identify (and when necessary, isolate) people with SARS. Scientists in Hong Kong demonstrated a way to detect the virus in respiratory aspirates (material taken from the lungs and bronchial passages) and in fecal samples.(2)

How does SARS spread?

The main way that SARS seems to spread is by close person-to-person contact. The virus is thought to be transmitted most readily *by respiratory droplets* produced when an infected person coughs or sneezes. Droplets can be spread when they are propelled through the air (generally up to about three feet) from a cough or sneeze. They can then get deposited on the mucous membranes of the mouth, nose or eyes of persons who are unprotected nearby. The virus also can spread when a person touches a surface or object contaminated with infectious droplets, and then touches his or her mouth, nose or eye(s).

It is therefore of paramount importance that with each index case, public health officials must trace all persons who come into 'close contact' with that infected individual. They must be interviewed, clinically assessed and placed under surveillance, if not isolation, in order to contain further spread of the SARS infection.

But that is only half the story. **It is also possible for the SARS virus to be spread more broadly through the air at least.** New research, published in the New England Journal of Medicine, debunked previous theories about the way SARS spread through the private Amoy Garden apartments in Hong Kong in 2003.(3) When more than 300 persons in the apartment complex fell ill, authorities first blamed it on sewage, roof rats and direct or indirect person-to-person contact. But the source of that outbreak was a person with diarrhea in building E. A team led by Ignatius Yu of the Chinese University of Hong Kong acknowledged that WHO investigators were partly right when they concluded that the illness initially spread within that building because an exhaust fan in the bathroom drew the virus up through traps in the floor drain. That was an interesting and coincidental scenario. Let's explain.

Normally, there is standing water in those drain traps. But strange as it may seem, many residents of that building had not been cleaning their bathrooms by flushing water down the drains. Instead, they had been damp mopping the floor, thereby spilling little water. Many traps had thus dried up and become ineffective in their essential function as such. Yu's group built a mock-up of the drainage system and demonstrated that huge numbers of aerosols were generated in the plumbing system when the toilets were flushed. They were then transported untrapped through the system. An exhaust fan in a bathroom drew the virus up, hitching a ride as it were, on microscopic airborne water droplets passing untrapped through the floor drain. The fan then sent the virus for an extensive ride, on these invisible aerosols, to other parts of the 36-storey building to do its dirty work.

But how did the disease jump to other buildings and why only *certain* other buildings were affected? The answer became obvious. Computer simulations of wind and airflow indicated which directions the virus might have traveled. Lo and behold, the people whose windows were downwind from the apartment where the first SARS case was reported, were 3-5 times more likely than others to fall ill. Powerful bathroom and kitchen exhaust fans in the complex must have drawn in contaminated air and helped to spread the SARS infection. Only a few tenants in 12 other buildings in less favorable wind directions, became sick.

Preliminary studies in some research laboratories suggest that the virus may survive in the environment for several days.(4) The actual length of time may depend on the type of material or body fluid containing the virus and various environmental conditions such as temperature and humidity.

Measures to contain SARS have taken two major forms: (i) isolation of symptomatic cases to prevent further transmission, and (ii) quarantine and close observation of asymptomatic contacts of each index case, so that they may be isolated as soon as they show any possible signs of the disease.

At the time of writing (February 2006) there is no known SARS transmission anywhere in the world. The most recent human cases of SARS-CoV infection were reported in China in April 2004 as an outbreak resulting from laboratory-acquired infections.

This has been a real wake-up call!

HYPERSENSITIVIY DISEASE

We have addressed some of the more important *infectious disease* consequences of indoor air contamination. That is one category of disease that has gained worldwide attention. At the same time, as was pointed out in Chapter 2, there is a clear correlation between the dramatic rise in the incidence of asthma in the population and the deterioration indoor air quality over the last three or four decades. The fact is that indoor air pollutants not only contribute to the spread of infection but they also severely affect those who show abnormal or *maladaptive responses* of the immune system to a substance recognized as 'foreign' to their particular bodies.

These 'hypersensitivity' responses are most obvious at the interfaces where the body encounters the airborne contaminants. The stereotypical response is seen in the reactivity of the airways where pollutants in the inhaled air interact with epithelial linings to trigger an inflammatory response leading to ***allergic asthma***. If the pollutants get deeper into the lung they interact with the tiny alveolar air sacs and again trigger an allergic response, this time labeled as ***allergic alveolitis*** (or 'hypersensitivity pneumonitis'). The symptoms can be confused since they typically are 'flu-like' with some fever, chills and associated chest tightness. Possible association of the symptoms with exposure to specific environments will often be the suggestive clue to a diagnosis. If the pollutants (particularly airborne pollen, dust mite allergens, animal dander and fungal proteins) provoke the classic 'runny nose', or nasal congestion, with or without sinus pain or tenderness, the response can then be labeled ***allergic rhinitis*** or sinusitis. These could be the most aggravating symptomatic conditions in susceptible individuals. A '*bad allergy*' day is a really bad day!

The other two interfaces besides the respiratory tract are: first, the sensitive conjunctiva that line the outer aspect of the orbits (eyes) and the eyelids, and secondly, the more robust and resilient skin surface. But here again, airborne pollutants can have observable consequences in either case.

Triggers for Bad Breathing

In the last chapter, we pointed out that airborne pollen can have debilitating consequences for susceptible individuals, especially during the so-called 'allergy season' in Spring and Fall. For those who suffer in this way, these are dreaded periods of the year. The widespread use of over-the-counter antihistamines and the prescription of steroids and allergy shots clearly illustrate the alarming prevalence of this condition in the general population. It cuts across all demographic lines. Any cost-effective solution that can significantly reduce the concentration of airborne pollen where people live and work, will clearly provide a major health benefit to society and could, in principle, dramatically ease the healthcare burden.

Allergic asthma is now being experienced in epidemic proportions. The rate of positive tests that measure sensitivity to fungal allergens, range from about 10% to as much as 60% of subjects tested. The rate observed depends on the particular population chosen, the allergen extract (challenge) used, and the method of testing. In the inner-city, about 40% of asthmatic children had a positive skin test to two common fungal extracts used. It now seems that sensitivity to fungi and subsequent exposure is a risk for dying from the complications of asthma. Allergic symptoms related to indoor fungal aerosols have been commonly reported, and there is clear correlation between symptoms of respiratory disease, including asthma, and airborne levels of specific fungi in homes.

Some sensitive people, when they breathe in fungal spores like the rest of us, go on to develop fungal colonies in their sinuses which result in further nasal inflammation and secretions. The sinuses become filled with a mass of fungal hyphae mixed with very thick secretions that can become tender and even painful.

Sight For Sore Eyes

One of the most common symptoms experienced in 'sick buildings' and particularly by individuals who show hypersensitivity and allergic responses to environmental contaminants is the irritation of the eyes. This may be described in many different ways: some would speak of "watering eyes", "itchy eyes", "burning eyes", "tired eyes", "sandy eyes", "stinging eyes", "painful eyes", "red eyes", "inflamed eyes", and even sometimes "dry eyes". But they all add up to discomfort, soreness and irritation. Among 'sick building' complainants, the prevalence can be as high as 50%.

The eye is protected by the immune system just like other mucosal surfaces. When it is damaged, invaded by pathogens (bacteria, viruses) or challenged by allergens (pollen, VOCs, dust etc), it mounts an immune response. White cells (neutrophils and lymphocytes) normally found in the tear film provide protection, as well as immune-active substances

When the eye becomes irritated by pollutants, it produces more tear fluid and it usually becomes somewhat red, due to the increase of blood vessels and blood flow to the conjunctiva. However, the number of visible vessels is also associated with infection itself, as well as with increasing age and blood pressure, so all the factors have to be taken into account. With the irritation there is often associated pain from the exposed receptors on the eye surface.

Normal house dust particles and normal non-reactive VOC's do not appear to produce significant eye irritation. The problem is with known allergens and reactive species. Hence the common symptoms are seen with pollen, pet dander and reactive organic species like formaldehyde and volatiles from cleaners and disinfectants, for example, in the home environment. The effects are exaggerated, of course, in individuals having conditions which modify their susceptibility, whether it be intrinsic like typical allergy sufferers, or extrinsic because of things like makeup use or cigarette smoking.

Irritants for Sensitive Skin

Just like the eye, the skin is also an interactive boundary between the human body and the environment, including the ambient air. It is crucial to human health, providing protection against radiation, mechanical injury, chemical contamination, and invasion by pathological microorganisms. It is also a regulator of body temperature and a shield against dehydration. But it can be adversely affected by pollutants in the air.

Complaints related to the skin are also quite common in cases of 'sick building syndrome' and the somewhat related condition of 'multiple chemical sensitivity.' Most often, patients complain of dryness and itching - a kind of general irritation and both allergic and contact dermatitis have been associated with some indoor air pollutants. In particular, nitrogen dioxide, pentachlorophenol, isocyanate, formaldehyde and other *VOC's* have been shown to cause irritation effects on the skin. These are associated with indoor air exposure when there is a time-based relation to specific microenvironments, especially if other co-inhabitants or coworkers have similar problems.

Dermatitis has been shown to result from exposure to *artificial fibers* released into indoor air from some construction materials. In such cases, the glass fibers penetrate directly into the skin and cause the irritation directly. On the other hand, *latex powder* from gloves can float around in the air of hospitals and can affect sensitive individuals by provoking allergic dermatitis. Exposure to allergens is even more relevant for the trigger proteins from pollens, animals, house dust mites and molds. They of course trigger respiratory symptoms but also provoke, in susceptible individuals, extrinsic forms of atopic eczema, which is a disease with inflammatory, chronic or relapsing skin lesions (rashes) and intense itching. As such, atopy is an apparent risk factor for the 'sick building syndrome' which we will describe in more detail next.

SICK BUILDING SYNDROME OR WHAT?

As we pointed out at the beginning of this chapter, the term 'sick building syndrome' (SBS) is used to describe situations in which building occupants experience acute health and comfort effects that appear to be linked to time spent in a particular building, but no specific illness or cause can be unambiguously identified. The complaints may be localized in a particular room or zone, or may be widespread throughout the building. In effect, it is what we call in medicine, a '***diagnosis of exclusion***'. After other classified illnesses are ruled out, and no obvious airborne contaminant is singularly suspected, then 'sick building syndrome' becomes the 'last resort' dumping ground to label the rather vague or nonspecific complaints of the victims.

In 1984, a WHO Committee Report suggested that **up to 30% of new and remodeled buildings worldwide may be the subject of excessive complaints related to indoor air quality**.(5) This condition is often temporary, but some buildings clearly have long-term problems. These problems frequently result because a building is operated or maintained in a manner that is inconsistent with its original design or prescribed operating procedures. Sometimes too, indoor air problems develop due to poor building design itself or because occupants engage in activities that exacerbate the contaminants of the local environment.

SBS does have some common characteristics. The WHO compiled a list of typical symptoms associated with acute discomfort.(6) They included:

- Eye, nose and throat irritation
- Sensitive, dry mucous membranes
- Dry, itching and red skin
- Headaches and mental fatigue
- High frequency of airway infections and cough
- Hoarseness and wheezing
- Nausea and dizziness
- Nonspecific hypersensitivity

Diagnosis of SBS would require a demonstration of an 'elevated' complaint or symptom prevalence associated with a particular building. The term 'elevated' remains vague, since there are no universal standards for 'normal' levels of any of these complaints. Complaints themselves reflect a range of factors like individual sensitization, economic and psychosocial parameters etc. But despite the vagueness of this characteristic, it is usually a starting point for investigation of a building, when the number and frequency of complaints or personal anecdotal experiences warrant an investigation.

Nevertheless, the cause of the symptoms is usually not identified even after thorough investigation. Yet most of the complainants will report relief of their symptoms soon after leaving the building. Studies do show that the reported symptoms are often caused or exacerbated by indoor air quality problems.(7)

The incidence of non-specific but recurrent health complaints among occupants of some buildings across the world is much more widespread than you might otherwise think. The undiagnosed illnesses can lead to personal morbidity, reduced productivity and increased absenteeism. We know that these health consequences are due to inadequate ventilation, chemical contaminants from a wide variety of indoor sources (mainly) and some that enter from outside, and biological contaminants including bacteria, molds, pollen and viruses. These elements may act singularly or in combination and may supplement other types of complaints, but in the case of SBS, the specific causes of the complaints will usually remain unknown, even after a thorough building investigation.

IT'S ALL ABOUT PEOPLE

There's a subtle disadvantage in the use of the term 'sick building syndrome'. It almost implies that it's the building that is sick. But in real terms, and obviously so, only people get sick - not the inanimate structures and chemicals in the environment. It is the health consequences to occupants of any building that truly matter. In other words, it's all about people.

Some individuals suffer from an enigmatic condition known as **multiple chemical sensitivity** (MCS) or multiple chemical intolerance. They represent about 2-6% of the general population who report extreme sensitivity to a plethora of exposures, ranging from new carpets and furnishings, to paint odors and cleaning agents, to off-gassing from office equipment, and much more. These chemically intolerant people report disabling symptoms when exposed to fragrances, tobacco smoke, diesel exhaust, as well as particular foods, medicines, alcoholic beverages, caffeine etc. Such intolerances sometimes predate difficulties experienced in the complaint building by years, even decades. Their symptoms are triggered by common, low-level chemical exposures that leave the vast majority of co-inhabitants unaffected in any significant way.

The typical scenario begins with the individual having a flu-like illness that seems to persist. Fatigue is the most prevalent complaint and it may present as a case of mononucleosis ('mono', the viral 'kissing' disease seen usually in young people) or chronigne fatigue syndrome (a diagnosis which is sometimes acquired in the process). But it is neither of these. The multisystem symptoms and a new onset of chemical, food and drug intolerances usually appear after an 'initiating' exposure event. That is, after some holiday from work or a vacation away from home, in some relatively clean environment (in the countryside, up the mountain or by the seashore), the symptoms flare up on return to home or work. This may then unmask an environmental trigger.

If the individual is not made aware by such 'removal from and return to' the environmental trigger, they may often claim that their illness traces back to some specific exposure event - again an 'initiating' event. This might be a chemical spill, repeated exposure to solvents, an application of pesticide, indoor air contaminants resulting from new construction or combustion products etc.

This then seems to open up a 'Pandora's box' of other triggers like we have mentioned and often, the symptoms can become unbearable. Different exposures will trigger different constellations of symptoms in different individuals. For each individual, a given exposure will tend to produce a consistent pattern of symptoms, which themselves will vary for each particular exposure.

In addition to the extreme fatigue, MCS sufferers will often report severe headaches, muscle pain, memory and concentration difficulties, various skin conditions, shortness of breath and/or a variety of gastrointestinal disturbances.

But 'normal' people do not experience these same symptoms under the same conditions at all. They don't even do so generally when exposed to much higher concentrations of the same chemical triggers. Moreover, the triggers are such structurally unrelated substances, and the symptoms themselves are so diverse and dissociated, that all the normal principles of toxicology and hypersensitivity (allergy) go out the window. Yet the patterns of response have been well-documented in a dozen countries at least, and among demographically mixed groups.[8]

The afflicted individuals deserve recognition, respect and a caring professional response

There is now general agreement that such an illness does exist.[9] It is otherwise called *environmental illness* and it is acknowledged generally that this is a real illness with a physical component and attendant suffering. It is not essentially 'psychogenic' or as the lay person would say, it is not 'all in the head'. This is so reminiscent of the history of 'chronic fatigue syndrome' and of 'fibromyalgia' where it seems that the patients were well out in front of the professionals. One might even go back further to recall the fight for the recognition of true menopausal symptoms and all that ensued from the initial fireworks.

This is a real illness. Results of several state surveys have suggested that **MCS could be one of the most prevalent, if not *the* most prevalent, chemically related illness in the United States.**[10] A surprising number of people believe they are sensitive to chemicals and made sick by some common exposures, and at levels known to be orders of magnitude below those producing symptoms in most other people. The homogeneity of responses across race/ethnicity, geography, education and marital status is compatible with physiologic response or with widespread apprehensions in regard to chemical exposure throughout society. In one of the U.S. EPA surveys that we referred to earlier, nearly one-third of federal office workers in mechanically ventilated federal office buildings considered themselves "especially sensitive" to one or more common chemical exposures.[11] Somewhere between 15 and 30 percent of the entire U.S. population, report being "especially" or "unusually" sensitive to certain chemicals. But it is believed that the real prevalence of MCS is only 2-6 percent overall.[8]

Indoor air pollutants seem to be able to both initiate the MCS condition in those susceptible individuals and to repeatedly trigger the disconcerting symptoms. Levels of volatile organic chemicals during or soon after remodeling or new construction prove to be orders of magnitude above their trigger thresholds. But even such people react differently. One person, for example, might become ill after an application of pesticide at home. That's their *initiating* event. *Later* they might have the most pronounced symptoms in their workplace as exposure to fragrances, new carpets, paint, furnishings, particle board etc. These would be unpredictable *trigger* events. In other words, the pesticide exposure may cause, in this case, subtle and gradual loss of tolerance, and the triggers at work produce vigorous and immediate symptom response.

That would be typical of more than half of MCS sufferers. However, about 40% of chemically intolerant individuals have no recollection of any *initiating* events. Some would describe a *series* of exposures that to them account for a more stepwise deterioration of their health. Yet others can't point to any specific changes like that at all, but rather, they just know that they have had unexplained lifelong health problems and intolerances that they find become exacerbated by indoor air contaminants.

All these people have experiences that correlate and despite the controversy surrounding the MCS phenomenon, the afflicted individuals deserve recognition, respect and a caring professional response that seeks to understand the true nature of the illness and to find ways to alleviate the condition that can at times become debilitating.

Today there is still no widely accepted medical treatment for MCS. A comprehensive approach might include the identification and avoidance wherever possible of chemicals and foods that trigger symptoms; cost-effective alterations of patients' physical environments with attempts at air purification when necessary, and appropriate social and psychological support. Put together, these interventions might help to alleviate the patient's suffering and delay or prevent worsening of their condition. But patients are still very much on their own. Often facing *unaware and skeptical healthcare providers*, they tend to drift from one to another in search of understanding and help.

Unfortunately, the best that can be said today is that full recovery from the MCS illness is seldom seen and that the condition is increasingly being recognized as a true disability. Some patients are still able to continue working, especially when appropriate adjustments have been made.

CANCER IS IN THE AIR

In contrast to the sick building syndrome and multiple chemical sensitivity responses, 'toxic reactions to indoor air pollutants' is perhaps the best understood category of all the health consequences because so much research has been done on this, though mostly in industrial environments. These reactions involve exposure to known contaminants such as carbon monoxide, lead or pesticides, which could result in acute disruptions of a variety of organ functions and/or an increased risk of chronic diseases, such as cancer. That will be the focus here, because there is cancer in the air.

Carcinogenicity (the ability to cause cancer) has now been classified into different schemes developed by national and international organizations for easy and agreeable identification of different carcinogenic substances. In the most useful scheme to date, substances are classified in groups as in **Table 2.**

Table 2. Classifying Carcinogens

Group	Class	Evidence
A	Human carcinogen	Sufficient evidence of causing cancer in *humans.*
B	Probable human carcinogen	Limited evidence of human carcinogenicity or sufficient evidence of causing cancer in *animals.*
C	Possible human carcinogen	Limited or inadequate data concerning an agent's carcinogenic potential.
D	Unclassified	No definitive evidence is known.
E	Noncarinogenic to humans	At least two adequate negative animal or human studies.

Acetaldehyde is a colorless volatile liquid or gas (at about 20°C) with a pungent fruity odor, widely present in the ambient work and indoor environments. It is formed as a major by-product of hydrocarbon oxidation when organic matter is combusted (burned). For example, it is produced when wood is burned for heating or cooking, or when there is incomplete combustion of fuels like gasoline, oil and kerosene. It is also a major constituent of automobile emissions and both mainstream and sidestream tobacco smoke.

Acetaldehyde is a probable carcinogen in humans. It causes the cross linking of DNA and more. In animal studies, acetaldehyde has caused squamous cell carcinomas and adenocarcinomas in the nasal cavities of rats and hamsters. There have been no similar studies in humans and just as well.

Acrolein is a colorless to yellow liquid with an unpleasant choking odor. It may be produced and released into the environment as a combustion/chemical oxidation product from the heating of oils and fats containing glycerol, wood combustion, cigarette smoke (especially sidestream smoke) and automobile and diesel exhaust. Significant exposure to acrolein might occur in workplace environments where it is used in manufacturing, as an end-use product, or as a result of combustion reactions.

Acrolein is a possible human carcinogen. It clearly induces DNA changes (even more than acetaldehyde) and it has been proven to be toxic to genes and cells *in vitro*. But it has not been definitively shown to cause cancer in animals or humans.

Asbestos has been described in the previous chapter. Among workers chronically exposed to asbestos, such as miners and insulators, there is an increased incidence of lung cancer (especially for cigarette smokers). *Mesothelioma* is the cancer associated with some cells lining body cavities and in the case of the lung, cancer of the pleural lining has been strongly associated with asbestos exposure.

Benzene was also mentioned earlier. At least half a dozen large-scale studies of personal or indoor air levels of benzene have been done since 1990. It has been identified in environmental tobacco smoke, which contaminates about half of all U.S. homes and workplaces, although that is showing a steady decline overall. It is elevated in the breath of smokers by a factor of 10 compared to nonsmokers. Some 50 million smokers in the U.S. are exposed to about half of the total nationwide "exposure budget" for benzene. The remainder is fairly evenly split between personal activities and the traditional outdoor sources.(12) But the traditional sources - industry and automobiles - actually contribute 99% of that benzene emission. In other words, smoking may be much less in real terms, but the exposure is relatively high because the pollutant is up close and personal, going right into the lungs.

Benzene is a known human carcinogen, with lots of evidence to support that conclusion. Even human epidemiologic studies demonstrate the influence of environmental levels of exposure on the risk of getting cancer. Children of smokers die at least twice as often as children of non-smokers, consistent with the high levels of benzene in the breath of smokers (which itself is suggestive of exposure in the womb). However, other data suggest a possible role for inhalation of airborne benzene.(13)

Fortunately, given the toxicity of benzene, its use has been dramatically reduced and the much safer toluene solvent has been a convenient substitute for many applications in manufacturing, especially of consumer products.

Cellulose was also mentioned before. Wood dust - also known as 'saw dust' and composed mainly of cellulose, polyoses, lignin and lower molecular weight ingredients - has been classified by the International Agency for Research on Cancer (IARC) as a known human carcinogen. Hardwood dust has been shown to cause adenocarcinoma of the nasal cavities and the paranasal sinuses.

Formaldehyde is among the top 10 organic chemical feedstocks (raw materials) in the United States. It is a colorless liquid or gas with a strong pungent odor and a tendency to rapid polymerization if not stabilized. It is available commercially in many different forms and is used in chemical manufacturing of a wide variety of resins for use as adhesives, textile treatments, finish coatings and insulation products - just to name a few. It is a product of organic material combustion as well as cigarette smoke, auto exhaust, wood fires etc. Formaldehyde is very common in the workplace, used by some 100,000 corporations with more than 2 million employees, many of whom have been exposed to it. It has also been demonstrated at elevated concentrations in many residential, institutional and commercial buildings, often at levels more than ten times the typical values outside in ambient air.(14) Most of it in these non-industrial indoor environments, originates from wood products with formaldehyde-resin adhesives and construction materials (particularly in insulation applications, before it was banned in the U.S. and Canada in the 1980's). Since that time, changes have been made to effectively reduce its use. **Smoking though, continues to be a local source of formaldehyde indoors.**

Formaldehyde causes many different health effects in several organ systems, from the respiratory and central nervous systems, to digestive and reproductive systems and more. Most importantly here, formaldehyde has been classified as a suspected (probable) human carcinogen based on proven facts. It is toxic to genes and DNA in vitro; it induces squamous cell carcinoma in the nasal/passages of laboratory animals, and shows some positive association with cancers of the upper respiratory system in humans, in epidemiologic studies.(15)

Pesticides are well known carcinogens. They are classified in different groups A, B, C as described earlier. The group A pesticides consist of inorganic arsenic and chromium compounds which are non-volatile and hardly airborne. But several pesticides used for residential and institutional pest-control indoors are classified in Group B (probable human carcinogens). Some are clearly volatile and a few have already been taken off the U.S. market. The level of pesticide residues (as can be measured by indoor air sampling), combine with the amount of residents' time spent indoors, the distribution in different rooms, and the levels of physical exertion, to determine the actual human exposure and the associated risk of deleterious health effects, including cancer.

Polyaromatic hydrocarbons have been mentioned in Chapter 3. They represent a large class of aromatic compounds found especially as products of incomplete combustion. They have been long known to be important contaminants of indoor air and many of them have also been recognized as human carcinogens. They are projected into indoor air from the burning of wood and hydrocarbon fuels, from unvented gas appliances, environmental tobacco smoke, furnaces, and from cooking, frying and grilling. They include molecules like benzpyrene, naphthalene, anthracene, phenols and phthalates, and many of them are associated with lung and bladder cancer, as well as skin cancer.

Radon, as we pointed out in the last chapter, has been known to be a cause of lung cancer for more than one hundred years. Today it remains a ubiquitous indoor air pollutant in many homes across the United States and as its natural 'daughter isotopes' decay spontaneously, they deliver the energy to target cells in the epithelial lining of the human respiratory system, to give rise to the radon-associated lung cancer. This, as you would expect, is only made worse by smoking. It is worth repeating that altogether, this accounts for about one third of all the lung cancer deaths in the United States each year.

Now we know the nature of the environmental **Problem** we have indoors. In a single phrase, **the air you breathe inside your home or office may be harmful to your health.** This raises concerns that you can't ignore, because there are **contaminants** in the indoor air environment that you can't avoid. To put it simply, it is not what you can see that threatens your health, but what you cannot see. Exposure to these indoor contaminants could lead to **consequences** you cannot afford. In the final analysis, many toxic effects that you don't see overnight, might be seen over a lifetime.

But you can do something. **There is a Solution to this Problem. Fresh Air for Life** is still possible! That Solution is the subject of Part Two of this book. But before we go on to discuss the **Solution**, it is important to deal with an issue that has become in recent time, a subject of much controversy and debate. In the field of indoor air quality, and particularly when it comes to air purification, there is nothing more controversial than the issue of ozone. This is understandable for the public is grossly misinformed on this subject. Ozone is indeed a paradox.

We'll explain this and more in the next chapter.

FIVE

The Ozone Paradox

"When the ***facts don't fit,*** *as in this case, you must acquit."*

In the field of air pollution, ozone does represent an interesting paradox. The *Webster's College Dictionary* defines a paradox as "*a person, thing or situation, exhibiting an apparently contradictory nature*". And that is certainly true of ozone. The ordinary *man in the street* has probably learned through the media and popular misinformation that ozone is something toxic in the air. It is supposed to be a real health hazard and major pollutant to be avoided. In fact, when most people think of air pollution, they often think of ozone levels and relate the two directly. Therefore, just the concept of ozone as a purifying agent would be an apparent contradiction in terms. But the contradiction is only 'apparent' **for there is in practice, no safer, effective and more convenient oxidizing agent (acting alone) to purify polluted air, than this very natural element!** Moreover, it is in part, nature's way of cleaning the air!

But first, we need to relate the facts ... just the straight facts about ozone.

OZONE 101

Ozone is a form of oxygen - the same elemental or atomic composition as the indispensable lifesaver in the air, that we all know - but it is a molecule that contains three atoms rather than the usual two. It is formed from the naturally-occurring, stable oxygen molecule by the addition of an extra oxygen atom. The result is a relatively unstable molecule that is willing, and even anxious, to give up that third atom of oxygen to other substances. That property makes it a powerful oxidizing agent.

Ozone is therefore a strong cleaning, purifying and oxidizing agent that reacts with organic molecules and substances to remove unpleasant odors, neutralize other toxic volatile compounds and kill germs. Paradoxically, this oxidation potential is its strength and at the same time, also its weakness. It represents both challenge and opportunity, if we would learn to capitalize on this unique property.

But that's the nature of things. The same sunlight that sustains all life on earth also causes fatal melanoma. The rain that waters all vegetation, can destroy cities by flooding, as Hurricane Katrina reminded us. So much for the **natural** *blessings* of sunshine and rain, they can become a curse. Think of a knife. What a remarkable invention. A knife in the hands of a skilled surgeon, is capable of many life-saving procedures and is so applied everyday around the globe. That same knife, in the hands of a serial killer, is likewise the agent of death and destruction. Or, think of a fire. There's another human discovery with the potential for sustaining and even saving life. However, when out of control, fire can burn relentlessly to destroy our forests or homes, and all the evidence of civilization anywhere. Even the precious gift of water can cause death by drowning, mud slides, property damage and disruption of public and private activities. Ingesting too much water at one time can be dangerous to one's health. Yet no one could deny the immeasurable value of water or fire. We would also have so much to lose if knives were outlawed or even worse, totally eliminated. Life could not be sustained at all if we removed all the water vapor or humidity from the atmosphere. (Perhaps that's why the NASA scientists were so jubilant when they recently found the first strong evidence of water having existed on the surface of Mars.) It is almost a truism, that things which have the potential for the greatest good, often (if not always) have a similar potential for harm - whenever misused or abused in inordinate ways.

In a similar fashion - despite its irritating effects on humans *in high concentration* - ozone has found widespread application in industry for many years. It has been used for water purification, odor control and as a disinfectant, just to name a few. The recent approval of the U.S. Food and Drug Administration (FDA) for the use of ozone to sterilize food and food contact surfaces, has opened up the door to many more exciting possibilities for this technology. In the field of air purification, we have asserted that there is nothing better than nature's own way.

Just as it does naturally outdoors, ozone does have great potential for improving indoor air quality under the right conditions. But before we get carried away, we should get back to basics. This section should be Ozone 101, after all, not Ozone 401.

Pure ozone is produced naturally and has a distinctly pale blue color when it is isolated as a crystalline solid or liquid at very low temperatures.

Typically, it is a colorless gas but in 'thick layers' it appears blue, which could in part, account for the usual blue appearance of the sky. When it is generated commercially from the air, a colorless gas is produced. At low concentrations, it has a very distinct sweet and pleasant odor - which has at times been referred to as the smell of 'fresh, clean air' as '*in the mountains*' and '*near waterfalls,*' or '*after a thunderstorm or lightning.*' Others compare it to '*the smell of spring*', especially after the rain, or yet to '*fresh, clean sheets.*' It gives air a quality of freshness that is universally appreciated. It is interesting that the name ozone was given to this gas because of its characteristic odor. The word is derived from the Greek word '*ozein*' which means 'to smell'.

Areas that are considered the healthiest vacation spots in the country have some of the highest levels of naturally occurring ozone. How about Waikiki Beach in Hawaii or the Padre Island National Seashore, the 100-mile long beach and dune region near Corpus Christi, Texas? Or would you rather go to the Grand Canyon in Arizona, the Glacier National Park in Montana or to the shores of Cape Cod in Massachusetts? These places are all known for fabulous vistas and clean, healthy air. The summer air in Arizona is hot and dry, while around the beaches it tends to be humid and windy. In northern Montana near the Canadian border, the sun is bright and the air is real crisp. All these places have natural ozone levels that go higher in the brighter daytime hours and lower during the night.

A protective layer of ozone is present far above us, in the region located 6 to 30 miles above the earth's surface at a concentration of about **1-10ppm** (parts per million). This layer is referred to as ***the stratosphere*** and performs the indispensable function of absorbing most of the harmful ultraviolet radiation, to shield the earth and its inhabitants from the sun. It also prevents heat loss from the earth's surface acting as a trap of infrared radiation below. Life on earth would not be possible without this. To the environmentalist, this ozone layer is of great concern and the effects of *either* increasing it to create the 'greenhouse effect,' or *else* depleting it to let more ultraviolet through to cause increased incidence of skin cancer and much more, are *both* midnight or even doomsday scenarios. These must be avoided at all costs, and outdoor air pollution poses a serious environmental threat. But that's for another place.

Natural ozone concentrations at ground level can vary between 0.01 - 0.05ppm (parts per million) or **10 -50ppb** (parts per billion), depending on the geographic location, altitude and season (local weather). It is formed spontaneously by the action of the sun's ultraviolet radiation on the copious oxygen component of the atmosphere with which it forms a stable but dynamic equilibrium. That is, the ozone molecule is unstable, with a half-life of about 20 minutes, so it is continuously being produced and then it reverts back to the more stable oxygen molecule.

Let's look at the equilibrium a slightly different way. Oxygen is naturally released from plants and sea plankton during photosynthesis. This oxygen diffuses or drifts and floats upwards into the upper atmosphere. It is transformed to some limited degree into ozone by the action of the sun's ultraviolet rays. The ozone that is formed is heavier than normal air, so it falls or is carried down draft, back into the lower atmosphere. But as it is 'falling,' the ozone molecules combine and oxidize any pollutant or contaminant they come into contact with, usually turning the latter into a safer product. This is a natural cycle, taking place all the time. In the process, the extra oxygen atom is released, with powerful oxidizing potential. With no new ozone generation, its concentration would be reduced to half its value in any given 20 minute period or so. The effect of atmospheric pollution is to shift this equilibrium so as to increase the transient ozone concentration.

OZONE HISTORY

Ozone was discovered in 1840 by a German chemist but its utilization in industrial situations has a rather long and impressive history. This was way before the current environmental concerns put it on the scientific center stage.

The American Indians loved to fish and depended economically on their catch.(1) They recognized a distinct correlation between a successful catch and a 'strange odor' released in the atmosphere by the action of lightning after an electric thunderstorm. Half way around the globe, the Greeks had also made similar observation and even defined the odor by the term '*ozein*', which we said earlier, translated means *to smell.* They too preferred to fish *after* a storm, and that is still common practice today. But little did either culture appreciate what was really happening after a storm. Apparently, ozone exerts a positive influence on the digestive system of different fish. It also has the ability to destroy viruses responsible for many diseases in fish culture. In any case, somehow the fish sense the ozone and are attracted.

The most common use of ozone to date is probably in the treatment of water for drinking. In 1906, a group of scientists and doctors studied the ozonation system at the Outshoorn plant in Holland and later constructed a plant in Nice, France capable of handling 19,000 cubic meters/per day, using ozonation for disinfection. For this reason, Nice is considered to be "the birthplace" of ozonation for drinking water treatment. Today that has become a widespread practice, in many parts of the world.

Ozone was first used on meats in 1909 in Cologne, France - ironically, by German scientists. The U.S. Department of Agriculture was slow to acknowledge and approve its use in foods but after many decades, they have now 'seen the light'! This has really been long in coming since an application of ozone use was first published in the United States in 1885. Dr. Charles Kenworthy of the Florida Medical Association reported on its use for medical purposes in Jacksonville, Florida. It has been utilized in the practice of alternative medicine in many other countries outside the United States for a long time. It is now widely used as an oxidizing, deodorizing, sterilizing and bleaching agent.(2)

Ozone is the second most powerful ***oxidizing*** agent in the world. It is second only to fluorine which itself finds very limited use in practice, since the latter is a rare and expensive halogen gas that has many devastating side effects and complications for routine human applications. Ozone is 150% stronger than chlorine. To be more rigorous and complete, hydroxyl radicals could also be mentioned ahead of ozone. But these unstable intermediates are only transient molecular species and do not form a characteristic substance that can be utilized as a reagent or other raw material like ozone, chlorine or fluorine.

Ozone is also the most powerful ***sterilizing*** agent, surpassing chlorine which is still in more widespread use. It works at killing bacteria more than 3000 times faster than chlorine. Ozone is effective in the destruction of bacteria, viruses, molds and odors and it does this with no other residue but the totally harmless oxygen as its only significant end product. This is an amazing characteristic. **As powerful an oxidizing and sterilizing agent as ozone is, the remarkable property is that it decomposes to a nontoxic, environmentally safe material.** How much better can a reagent get? This is in clear contrast to chlorine which produces hypochlorous and hydrochloric acid. Even in its widespread use in water, chlorine has complications of odor, taste, allergies and irritation for sensitive individuals.

In considering the history of ozone, perhaps it is appropriate to include a reference to a series of studies published at the height of the Cold War in the Journal *Priroda* (1976) where the Department of Health in Russia established some important points concerning the use of ozone in closed indoor environments.(3) They found that normal air loses its basic "freshness" quality when it was drawn into air conditioning and heating systems, with a loss of as much as 90% of its ozone and ion levels. This was demonstrated to cause symptoms for building occupants who complained of headaches, weakness and a general poor feeling, not unlike the 'sick building syndrome' described more fully in the last chapter. However, after five months of testing, with both a test group and a control group, the subsequent exposure to 15ppb (parts per billion, a level not so atypical of normal air) of ozone, restored a feeling of well being. Moreover, at these levels of exposure, they were able to observe increased immune potential, higher oxygen content in the blood, improved blood pressure readings, and the reduction of many stress characteristics associated with working in modern office environments.

The Russians went a bit further. They found that by reactivating the air with the 15ppb of ozone, the effect was comparable to that of taking an outdoor walk for 2 hours during the day. They took these expectations to schools. Studies by the Institute of Child and Adolescent Hygiene showed that by raising the ozone levels in classrooms to 15ppb, up to 69% of the students decreased the time required to complete concentration tasks. They also noted favorable changes in respiration, mental reserve capacities and general state of physical and mental health.

Now, there's something to think about.

OZONE FORMATION

In unpolluted areas, ozone is created by the action of **ultraviolet radiation from the sun** acting on naturally occurring nitrogen oxides, methane molecules from normal agricultural and animal husbandry, and even the hydrocarbon compounds of isoporene and terpene emitted from trees. Anywhere that hydrocarbons exist in nature with the presence of strong sunlight and moisture, there will be some degree of ozone formation. That will vary with local conditions.

In the uncommon but usual **thunderstorm and associated lightning**, ozone will be generated more efficiently. The electrical discharge triggers the partial dissociation of the normal oxygen 'diatomic' molecules (Step 1). The individual atoms can subsequently attack and attach to other oxygen molecules to form the ozone 'triatomic' molecules (Step 2). This again accounts for that sweet pleasant odor of 'clean, fresh air' following these storms.

Fig. 1 Ozone Formation

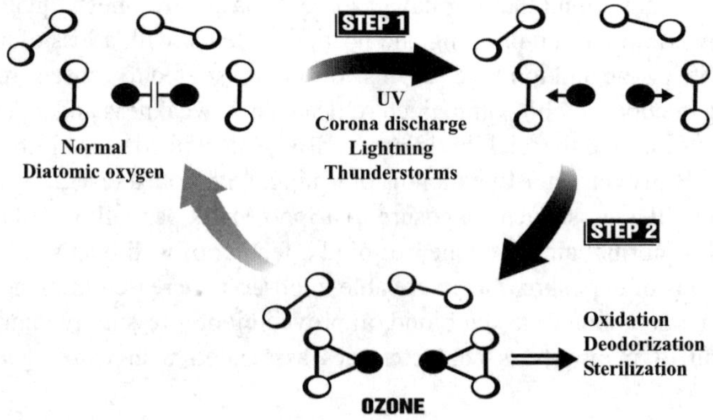

In urban areas, ozone is also created in two further important ways. In the first place, the **effluents of industrial processes** consist of a variety of toxic and noxious chemicals that contaminate the environment in ways that now provoke the average citizen. The sight of chimneys and smoke-stacks with billowing white or black clouds above, creates chilling implications for future generations. They signal the rising concentrations of 'greenhouse gases' in the atmosphere. These chemicals in the presence of oxygen and ultraviolet rays also give rise to ozone production.

Secondly, **automobile emissions and mass-fuel burners** likewise produce hydrocarbons and other pollutants that, in the presence of oxygen and the sun's ultraviolet, lead to photochemical production of ozone as well. The highest concentrations of ozone will be found in those areas with the highest concentration of unoxidized or unburned hydrocarbons. But remember the real *cause* and the true *effect*. Or in other metaphorical terms, think about which is *the chicken* and which is *the egg*. As we pointed out earlier, ozone is both a '*thermometer*' registering the extent of pollutants in the air, as well as a useful '*thermostat*' controlling the fate of those same pollutants by its oxidizing (cleansing) action.

That brings us back to the question of '*who's the real culprit* when it comes to atmospheric pollution?' It is easy to put ozone on trial, but if we quickly rush to judgment, we would make a big mistake. As the late Johnny Cochran would probably say, 'if the facts don't fit, as *in this case*, you must acquit.'

So let's go to trial.

OZONE ACCUSED

The 'Ozone Story' is often *confusing* and perhaps just as often, *misleading.*

First, let's just comment on the 'confusion'.

Indeed, ozone itself is perhaps the most misunderstood and confusing element in the air we breathe. It is also the most hated by some parties with limited knowledge or vested interests, while being at the same time, the most loved by many who benefit from its effective application in environmental clean-up. On one hand, we are told that it is a harmful, poisonous gas capable of seriously damaging our lungs. On the other hand, we are also told that it has the potential of being the greatest natural purification element to be found anywhere for dealing with the alarming levels of man-made pollution.

If you listen to weather and environmental reports (and who doesn't!), you would notice that some days you hear that ozone levels are too high and dangerous, whereas on other days, you hear that we are in danger of 'depleting the ozone layer' and we need to be more responsible and do something about that. Just to be helpful to the reader who is still confused - we point out again that 'the ozone layer' in the upper atmosphere above the clouds (recall '**the stratosphere**') provides a protective filter and shield for life on earth, by protecting us from the devastating effect of excess ultraviolet radiation from the sun.

That naturally occurring ozone layer is not to be confused with the artificial, man-made production of elevated ozone as we just described in the ambient air at ground level ('**the troposphere**') which we breathe continuously. That distinction is most important.

Ozone Reports can not only be 'confusing'. But at times can also be 'misleading', although not necessarily by intention. If you listen or read the consumer reports you would be inclined to think that ozone is the *only* 'pollutant' in all the smog and haze of air pollution. You would also imagine that ozone is *always* harmful to health. Neither of these ideas would be true.

It's like blaming the firefighters for the fire, while the city burns.

'Smog pollution' is by no means synonymous with 'ozone pollution.' Far from it! All the ingredients of smog are harmful in their own right, whether it be the volatile oxides of nitrogen and sulfur, carbon monoxide, hydrocarbons like benzene and polycyclic aromatics, or all the VOC's that we should learn to fear. Ozone seems like such an easy target, maybe because of its characteristic odor, its simple formula or its ease of measurement. But it is by no means, the only hazard in 'smog'. **However, the ozone level has become a convenient index of air quality (pollution) and is in widespread use. That's fair, except that it's an indicator of other things too, especially of how much we are polluting the atmosphere.**

As long as we conveniently place the blame on ozone as if it were the culprit - Public Enemy #1 - the government and the barons of industry are effectively relieved of the responsibility for the millions of pounds of toxins released into the air each year. That is at least the perception of some. Ozone is nature's doing, after all. But to think that's where "the pollution" is, would be a gross deception, bordering on folly. It's like blaming the messenger more than the message and trying to escape its demands. To change the metaphor, it's like blaming the firefighters for the fire, while the city burns.

Where this becomes truly misleading is in the description of health effects associated with ozone. Unqualified, ozone can be harmful. There's no doubt about that. But one must always qualify any statement about health consequences with some indication of levels of concentration. Failure to do so, makes almost anything conceivably dangerous to health, even the indispensable oxygen we breathe, the water we drink and any food we eat. Too much of anything is bad for you! That clearly includes ozone.

The U.S. EPA has identified a rating system for the health concerns associated with different ozone levels, based on an Air Quality Index (**Table 3**).

Table 3. Air Quality Index

Ozone Level (ppm)	**Effects**	**Color Code**
0.000 - 0.064	Good	Green
0.065 - 0.084	Moderate	Yellow
0.085 - 0.104	Unhealthy for Sensitive Groups	Orange
0.105 - 0.124	Unhealthy	Red
0.125 - 0.374	Very unhealthy	Purple

When the concentration of ozone is too high, ozone can be quite harmful. Again, some people are more susceptible than others, especially children, seniors, people who work all day outdoors, people with pre-existing respiratory conditions and those "responders" who are otherwise healthy but tend to have an exaggerated reaction to ozone.

In 2000, a study supported by the California Air Resources Board (under the auspices of the Long-Term Exposure Health Effects Research Program) regarding the effects of air pollution on children's lung function growth, determined that "**Ozone did not appear to play a major role in the pollution's effects on children's lungs. Instead, the offenders were nitrogen dioxide, microscopic particles known as particulate matter, and acid vapors**. All come directly or indirectly from the burning of fossil fuels (the exhaust from automobiles, for example), as well as from emissions from industrial plants and other sources".(4) Many other experts have unjustifiably and incorrectly singled out ozone for criticism for over fifty years, perhaps again only because ozone is so easy to measure and too easy to blame.

Some areas in the United States with elevated pollution levels, especially during the summer months, do produce enough ground-level ozone, to cause some immediate health problems. These might include shortness of breath, chest pain with deep inspiration, wheezing and repeated coughing, and increased susceptibility to respiratory infections. More chronic exposure to ozone can increase the risk of premature mortality, pulmonary inflammation and fibrosis, the risk of asthma attacks, and the need for medical treatment and hospitalization for asthmatics.

The particular effect of ozone on any given individual (and that includes you) will naturally depend on different factors such as personal susceptibility, the ozone concentration level, the rate of breathing and obviously, the length of exposure.

But note again, that ozone is the end result here - not the cause. Nevertheless, it is indeed unfortunate that smog and ozone have been too frequently interchanged in popular discussion of air pollution. That has masked the clearly positive characteristics of ozone as the natural way of dealing with air quality problems. The misplaced focus on 'smog' as 'the air pollution' has essentially prevented the general population from seeing the invisible - but even greater - problem of indoor air pollution, where ozone proves to be an even more effective solution.

However, despite all that you may have heard about the hazards of ozone (really, excess ozone), you will discover in this book that there is a wealth of data to demonstrate the valuable benefits of ozone in controlled air environments where levels of human exposure are relatively low. May it suffice to suggest for now, that the wise injunction **'don't throw out the baby with the bath water'** is as appropriate here as it has ever been! More on that later.

OZONE POWER

When used properly and safely, ozone technology can be an inexpensive and effective tool for eliminating many unwanted odors and pollutants present in indoor air. That subject will be covered in a later chapter. But some of the most successful applications of ozone have been for years now, in water treatment and in manufacturing industries such as food, beverage and pharmaceuticals, as well as in the healthcare and hospitality industries. **In the area of public health, ozone technology has been a potential source for reducing the role of infection, both in the home and in healthcare facilities.**

Ozone is widely used in the HVAC industry for duct cleaning and disinfection, where it is very effective against mold, mildew, odors and bacteria, including the organisms responsible for Legionnaire's disease. Municipal sewage plants use ozone to destroy sewer gases and other odors. The hotel industry has recognized its benefits as a room sanitizer, and its ability to permanently remove smoke odors. The food industry also utilizes ozone for mold, mildew fungi and bacteria control on food products. All these uses of ozone have one thing in common: they are commercial uses requiring operator training and caution with excessive dosages of ozone.

The complete range of commercial applications for ozone and new emerging technologies being developed would number perhaps in the hundreds.

Here is a short list of the principal applications for ozonation systems (since single ozone generators are hardly ever used as such):

Air treatment	Fish canneries	Water processing
Agriculture	Food sanitation	Pools and spas
Bottling water	Industrial waste	Odor Control
Cooling towers	Therapeutic use	Waste water treatment

We'll now discuss two of these very important applications, just to familiarize you with the useful application of ozone on a daily basis. The first application for the treatment of drinking water and waste water management has a long history, going back to the earliest uses for ozone in Europe at the turn of the last century. The second application for food treatment can also be traced back to 1909 when it was first used on meats. But it has taken almost a century here in the United States to gain FDA approval for the use of ozone as a sanitizer for food contact surfaces, as well as for direct application on food surfaces, as an alternative to chlorine.

Water Treatment

The safety of drinking water is of vital importance to the health of the general public. Protecting the source of the water supply to any community is of paramount importance when it comes to obtaining microbiologically safe drinking water. This must be a priority for the authorities everywhere in an age of global terrorism when criminal elements devise schemes aimed at inflicting the maximum disruption of the economy and causing multiple loss of innocent lives. So many diseases can be quickly and widely disseminated by water contamination. Indeed, many sources are normally highly polluted and need extensive treatment and sterilization before distribution to the trusting consumer.

Chemical disinfection is an important factor in water treatment systems throughout the industrialized world. Oxidizing chemicals like chlorine and ozone destroy a wide variety of disease-causing microorganisms during treatment. In many countries, chlorine is applied as an additional safeguard in the distribution system itself.

There are now hundreds of ozone water treatment plants across the United States, and thousands worldwide. Los Angeles reportedly has built the largest ozone purification plant in the world. More than 400 municipal water treatment plants are using ozone across the country. This ozone treatment (ozonation) is now easier, more efficient and much less costly and is gaining wider acceptance because it works, and works well, at an affordable price.

Ozone was approved by the U.S. FDA in 1982 for purifying bottled water and today in California, it is the only agent approved for that purpose. By

1991, the same department approved ozone for the recycling of Poultry Chill Water. Since 2001, ozone has been approved for sterilizing food surfaces and also for use in water to wash foods.

Ozone has many distinct advantages. It is not only a very powerful oxidizing agent but also a very effective non-chemical disinfectant. That latter term is appropriately used because, as we mentioned earlier, it is unique in decomposing to a harmless, nontoxic, environmentally safe end-product, namely oxygen. In Europe, ozone use is more widespread and finds chosen application for many purposes: color removal, taste and odor removal, turbidity reduction, removal of organics, microflocculation, iron and manganese oxidation, and most commonly, bacterial disinfection and viral inactivation. It is often seen as a preferred method for virus inactivation and other effects, rather than just a simple alternative to the use of chlorine.

The United States started chlorinating the water supply around the turn of the last century. At the same time, much of Europe decided to use ozone to purify its water supply. Many believe the U.S. went with chlorine due to a strong chemical lobby. Could this be true? We have known for many decades that chlorine reacts with organics and produces known carcinogens. Yet we continue to clean most of our water and much of our food with chlorine. High levels of chlorine are regularly used to clean vegetables, fruit, poultry and even grain.

Now, why did our nation's water suppliers and food processors continue to use chlorine even though we've known for decades that it causes cancer? Perhaps one could argue that it is far more dangerous to eat chicken, for example, contaminated with *salmonella*, or vegetables laced with *E.coli*, than to incur the finite risk of cancer associated with chlorine.

However, could there be a better way? The answer is a resounding Yes! The Europeans started using it in the late 1800's and the track record of ozone as a purifier is second to none.

Cryptosporidium parvum is the leading cause of persistent diarrhea in developing countries and is a major threat to the U.S. water supply. It has caused major outbreaks of waterborne diseases in Europe and in North America. It is sometimes found in untreated surface water, as well as in swimming and wading pools, daycare centers and hospitals. The *C. parvum* organism can cause illnesses lasting longer than 1-2 weeks in previously healthy persons or indefinitely in immunocompromised patients. Adequate control of this pathogen is therefore of critical importance in most surface water supplies. Water from storage reservoirs may contain similar concentrations of other pathogenic microorganisms such as viruses and *Campylobacter jejuni*. However, these organisms are inactivated by post disinfection processes such as UV irradiation and are therefore lesser sources of pubic health problems.

The use of ozone in water treatment has been a big success. **Indeed, ozone has proved to be an excellent disinfectant for the treatment of both water and waste water management,**(5) especially because it is the most effective against microorganisms. For example, *C. parvum* co-cysts are resistant to chlorination but are effectively inactivated by ozonation.(6) Ozone is also effective in reducing populations of coliform bacteria like *E. coli* and *Pseudomonas aeruginosa*, by more than 99% in waste water.(5) It is an excellent disinfectant for destroying spore-forming bacteria and viruses which are destroyed almost instantly.(7)

Ozone is also very reactive against sulfides in water, thereby effectively removing much of water's unpleasant odor. In fact, the oxidizing action can be broken down into two broad categories. Simple oxidation pertains to sulfur, iron and manganese which oxidize quickly and easily. They require relatively little ozone and show no detectable ozone residual. More complex oxidation pertains to the oxidation of contaminants such as volatile organics, organic odors, microbes such as bacteria, protozoa, amoebae and viruses. This requires more ozone and necessitates some ozone residual.

The clear bottom line is that protecting the municipal water supply is a critical consideration in any civilized society. It is essential for public health. Ozone is a proven safe and effective contributor to that end.

The water supply is no more vital than the food supply. The challenge to public health professionals is even more complex in the latter case.

Food Treatment

The food industries face a large number of issues when it comes to producing and delivering a safe, wholesome product to the consumer. Food pathogens such as *E. coli* 0157:H7, *Salmonella* species, *Listeria monocytogenes* and *Clostridium botulinum* have proved to be a growing concern over the years. Food processors have also been concerned about spoilage due to microorganisms which shorten shelf-life and cost companies millions of dollars every year in spoiled product.

This impacts many industries like meat, seafood, poultry, produce, baking, canned foods, dairy and almost every other segment of the market. The U.S. Department of Agriculture estimates the costs associated with food borne diseases in America to be about $5 billion to $22 billion a year, despite all the existing protection measures. That does not include the billions lost each year due to spoiled product which must be discarded or sold at huge discount. Better disinfection and microbiological control measures are needed in almost every area of the food industry as a whole.

Ozone is a powerful, broad-spectrum antimicrobial agent that has been found to be effective against bacteria, fungi, viruses, protozoa, bacterial and fungal spores.[6] Potential applications for ozone in the food industry might include:

- Increasing the yield of certain crops
- Protecting raw agricultural commodities during storage and transit
- Sanitizing packaging materials for storage and
- Adding to water for washing food in the preliminary and final stages of distribution.

In the United States, chlorine has been the standard disinfectant in many food applications for years. It is common in meat processing, for example, and has proven to be effective and safe when used at proper concentrations. However, chlorine is far less effective than ozone as an oxidizing agent and it is much slower in killing off bacteria and viruses. More importantly, it can react with meat to form highly toxic and carcinogenic compounds called trihalomethanes (THMs). These THM products have been implicated in cancers of the kidney, bladder and colon. Use of chlorine also results in the production of chloroform, carbon tetrachloride and chloromethane. On the other hand, ozone does not even leave a trace of residual product after its oxidative reaction is complete.

Meat is just one example. Fresh cut produce is a rapidly growing segment of the food industry. In 1999, it reached $76 billion in sales volume. But outbreaks of disease from food pathogens on produce have also increased, going from 4 per year in the 1970's to more than 10 in the 1990s. And that's only what is reported. Ozone and chlorine rinses again represent the two most common disinfecting treatments available in that area. Ozone was approved by the U.S. FDA for use as an antimicrobial treatment on produce in 2001 and it has certainly proved to be extremely effective. An ozone rinse of just 1.3ppm for 5 minutes produced a greater than 99.9 percent reduction in bacteria on lettuce.[8]

Just to mention one more class of foods, we might refer to the application of ozone in the flour milling industry. There it has been found to be an effective means of cleaning the grain in the mills' cleaning houses. One study performed at the Harvest Milling flour mill in Huron, Ohio showed a 75-80% reduction in total plate count bacteria in ozone treated flour compared to conventional treatment with chlorine.[9]

Ozone has some distinct advantages when compared to chlorine. It is clearly a better choice for disinfection of surfaces. That's because chlorine is halogen-based and corrosive to stainless steel and other metals used in food-processing equipment. It tends to form toxic gases when mixed with even small quantities of ammonia or acid cleaners. Ozone leaves no residual products in food and after ozonation, the foods can still be legitimately called 'organic'. Any organic sanitizer must be registered as a 'food contact surface sanitizer' with the

U.S. EPA. Ozone certainly is, and as of June 2001, **ozone has gained FDA approval as a sanitizer not only for food contact surfaces but also for direct application on food.**

The U.S. FDA made the recommendation to industry in 2004, that "Ozone is a substance that can reduce levels of harmful microorganisms, including pathogenic *E. coli* strains and *Cryptosporidium*, in juice. Ozone is approved as a food additive that may be safely used as an antimicrobial agent in the treatment, storage and processing of certain foods under the conditions of use prescribed in 21 CFR 173.368."

OZONE SAFETY

Every chemical substance has an effect on biological systems that can range from little effect on one extreme, to levels of lethality that simply means it can kill you, at the other extreme. In reality, every chemical has the capacity to be toxic and there are three determining factors of major importance:

- What is its concentration or dosage?
- What is the extent (duration) of exposure?
- Is there any predisposing susceptibility?

The case of ozone is no exception.

What happens if the concentration of ozone is increased in the surrounding air? *First of all*, the odor in the air changes from being characteristically sweet and pleasant, to a more pungent, irritating, 'bleaching' odor that alerts any conscious individual of the need for action. That action would necessitate removing the source of ozone - any typical source of ultraviolet such as lamps in photocopiers and other office machines, sunlamps, germicidal lamps, laboratory instruments etc. The operation of certain types of industrial machines and electrical equipment can also inadvertently produce elevated levels of ozone. The recurrence of that irritating odor should alert the individual to turn off the suspect equipment at once. *Secondly*, it is time to increase ventilation - that calls for opening windows or doors, or turning on a fan where appropriate. *Thirdly*, and most importantly, it is time to evacuate - to move away, to get outside and breathe some more normal air to purge the lungs.

But there is never a need to panic. Why? Because ozone is essentially safe, in almost all normal situations and even in practical ozone applications. **Table 4** shows the current standards for ozone accumulation levels and ambient outdoor levels.

Table 4. Ozone Safety Levels

Agency	Recommendation	Comments
U.S. EPA	$0_3<0.064$ ppm (parts per million)	Generally safe
	< 8 hr. exposure if $0_3 > 0.085$ ppm	For sensitive groups
U.S. Dept. of Agric.	$0_3 < 0.100$ppm	For continual exposure
U.S. FDA	< 0.050 ppm	For continual exposure
ASHRAE	< 0.050 ppm	For continual exposure
OSHA	8 hr exposure limit for,	For Industrial Workers
ACGIH*	Time-weighted average $0_3 \sim 0.100$ ppm	
ALA	For 8 hr exposure, $0_3 < 0.050$ ppm	UL standard for medical devices
Compare:		
Nature	$0_3 \sim 0.010$ - 0.020 ppm	Typical outdoor air levels
	$0_3 \sim 0.010$ - 0.030 ppm	Clear spring day in the mountains
Space-Age Technology Air Purifiers	$0_3 \sim 0.02$ - 0.04 ppm	In normal applications

CONSENSUS: SAFETY ZONE = 0.010 - 0.050 ppm

* Amer. Conf. of Government Industrial Hygienists

Despite these suggested guidelines (and in some cases, regulations), the fact is that no one has really demonstrated any serious or irreversible medical complications resulting from ozone exposure in normal operating environments. Because of the irritating odor which is sensed and responded to by normal individuals, well below all the standard limits proposed above, there has been no significant and substantiated reports of serious health consequences from ozone exposure in any typical working or living environment. Naturally, at very high concentrations '*it can be harmful*', but that could be said of almost any known chemical entity. Prolonged exposure to moderately elevated levels (probably above 0.100ppm or 100ppb) produces headaches and sometimes nausea. Emergency First Aid procedure is to simply remove the affected individual from any areas of high ozone concentration. On doing so, the symptoms will typically disappear in short order and without any serious consequences.

The question of safety must always be addressed when using ozone technology for indoor applications with possible human exposure. An earlier study by Boeniger [(10)] found that *ozone air cleaners* did pose a potential health risk if and when used at high levels indoors. However, responsible manufacturers who implement this ozone technology are well aware of this health risk and have worked to improve the science to make ozone much safer for indoor use. As you will learn later, **the best air purification technologies today can utilize ozone as an oxidizing, sterilizing and deodorizing agent without significantly increasing the levels of ozone in the air itself.** That's space-age technology and you will just have to wait.

In the application of any technology, the crucial question always revolves around what is termed *risk-benefit* analysis. In medicine for example, this is the daily mantra '*above all, do no harm*'. But every intervention has an attendant risk - a price to pay. Wise doctors are always weighing the two things in balance. '*What if we do this - what's the benefit to the patient and what is the risk?*' On the other hand, '*what if we don't do this - what's the cost to the patient and what is the risk in that case*?'

Aspirin is probably the most widely used drug of all time. It is like a miracle drug, with new benefits being discovered almost every few years. Yet, every day aspirin causes serious complications and deaths due to hemorrhage, anaphylaxis, Reye's Syndrome, ulcer exacerbation, and the list goes on. In fact, it has been vigorously argued that if aspirin were being introduced today, it would not be sold 'over the counter' because of its side effects and levels of risk. But the benefits of aspirin are now so well established that the risk-benefit analysis works in its favor. And society accepts the consequences.

The benefits of ozone as an oxidizing agent - that sterilizes, cleanses, deodorizes, neutralizes and in a word, decontaminates air, food and water - are obvious. The risk is the finite chance of misuse or abuse (by some uninformed or careless individual) that leads to elevated concentration levels which can produce mild to moderate and usually reversible effects (at worst). With careful consideration, the net balance in favor of cautious ozone use, for any proven application, is eminently obvious! It is the rational choice - both commercially and domestically.

Professor James Marsden of Kansas State University has studied this field for a long time and he sums it up like this:

> *"A study that states we should not intentionally breathe ozone is probably correct. But does this mean that we should do away with all ozone devices. Surely not. Chlorine, for example, has thousands of research papers written about it. The ill effects of chlorine have killed thousands of people through misuse and possibly millions from delayed cancer. However, it has saved hundreds of millions of people worldwide by its ability to kill water borne diseases. Can you find a study that says chlorine kills? Sure, but you can also find many studies that tout its virtues. Now, of course we are rapidly phasing out chlorine and replacing it with ozone that is far safer. But can ozone hurt us? Sure, too much of almost anything can hurt you. Remember, even too much oxygen can kill us."*[11]

Finally, on the subject of ozone safety, the author is indebted to Gil Nissen for an article she fabricated that to her mind, may well have appeared in an 18th century issue of Confusors Retort magazine.

> *"A young man has recently discovered electricity and is proposing that his discovery may one day be brought into homes across these great colonies. He proposes that electricity may be harnessed and transported into the home via copper wiring and that it will power grain grinders, electric mixing spoons, and maybe even lighting.*
>
> *"Confusors Retort has investigated the harmful results of placing such a deadly force into the homes of New England. It is confirmed that electricity can be generated with various technologies. Although these machines are being sold for people to bring electricity into their homes, there is no benefit to having electricity in the quantity of a mere 60 watts.*
>
> *"Placing electricity into homes will expose users to life threatening electrocution in many instances. If this expensive*

technology that the general store is offering is purchased by the unsuspecting public, thousands of deaths could result over future generations from children placing paper clips into the wall sockets. Other results of this dangerous product will be severe burns, damage to the eyes can result if it is applied for arcing metals together to weld iron. There will certainly be thousands of others who will severely burn themselves on hot surfaces heated by this naturally occurring phenomena if it is produced unnaturally and brought into our homes. Electrical shortages will cause millions of fires across our colonies and there will be untold losses due to fires that can be ignited by 'electrical wiring'... blah...blah....blah..."

The Moral to the Story: *A little truth mixed with a little ignorance can sometimes offer a lot of ignorance...not to mention* (by extrapolation) *the lack of electric light bulbs in America today*!

It was indeed a wise observation that 'a little knowledge is a dangerous thing'. It can sometimes stifle useful innovation, obscure real benefit and derail a good opportunity.

OZONE MYTHOLOGY

There is so much common misunderstanding about ozone and its use as a purifying agent. Therefore, even at the risk of being repetitive, we conclude this chapter by simply emphasizing again just five common myths in this area. The ozone paradox is such fertile soil from which mythology can grow.

Myth #1 **All air cleaners sold for residential use produce hazardous ozone and are therefore inherently dangerous.**

Nothing could be further from the truth. As we pointed out earlier (Table 4) all reputable bodies including the U.S. EPA, the U.S. FDA, the Occupational Safety and Health Administration (OSHA), the American Society for Heating, Refrigeration and Air Conditioning (ASHRAE) and the American Council for Government Industrial Hygienists (ACGIH) have acknowledged standards for ozone accumulation and exposure levels. Everyone agrees that at least up to levels of 0.05 ppm, ozone poses no significant health threat to the public. These are conservative standards.

The fact is that many reputable manufacturers of air cleaning devices have gone to great lengths to ensure that ozone production by these units does not exceed this standard. Even the U.S. EPA found that some ozone generators tested under manufacturers' recommended conditions, produced ozone at rates normally within the ranges stated by the manufacturers. When several respectable air cleaning devices are used properly and safely, there seems to be little concern for any serious health risk.

Millions of ozone air purifiers have been sold in the United States over the past few decades, but there have been very few, if any cases, where an ozone air cleaner has been definitely linked to any major illness. As we have already pointed out, in June 2001, the U.S. Food and Drug Administration (FDA) formally approved the use of ozone in gaseous (air) and aqueous (water) phases as an antimicrobial agent in food, including meat and poultry.

Myth #2 **Low-level altitude pollution or smog is due to ozone.**

In the age of sound-bites and live reporting, this is an understandable myth. It is true that wherever smog levels are high, so are the measured ozone levels. And while ozone is easy to measure, hydrocarbons are not. They are just too complex. In addition, ozone is always present in levels consistent with the hydrocarbon (pollution) level, therefore the common assumption is perpetuated that ozone is the culprit.

But it's just not so.

Again, it bears repetition: ozone forms naturally when sunlight reacts with man-made hydrocarbons in the air (generated by automobile exhaust and smokestack emissions, for example). Hence, **the more hydrocarbons there are, the more ozone is produced.** Moreover, it is the ozone produced naturally outdoors that is actually nature's secret for helping to break down those harmful hydrocarbons and other pollutants.

Therefore, to repeat the analogy (for it is a good one!), we should not blame the firefighters for the fire! Of course, the bigger the fire, the more firefighters are on the scene! So it's possible to measure the fire by the number of firefighters and fire trucks that respond. But the fire is always the real problem, not the emergency response. In reality, ozone is protective, not destructive. Without ozone, we would not be able to live in our modern polluted cities!

Myth #3 **Ozone is the toxic agent in polluted air.**

Many people believe that the offensive odors, burning eyes and aggravated respiratory problems associated with air pollution, are all due to high levels of ozone in the common polluted air. The fact is that they mistake hydrocarbons, oxides of carbon, sulfur and nitrogen, halogenated by-products, lead and other sulfur compounds - all these irritants and more - for a single scapegoat:

ozone. These other pollutants do indeed cause eye, nose and throat irritations. In their absence, the irritating effect of ozone at the low concentration levels is a far cry from what is commonly assumed.

It's like giving a dog a bad name, then the dog is always to blame. And then to go on and punish the dog for what it did not do, is a miscarriage of justice, plain and simple. Dogs are not the only creatures capable of mischief and quite often, little 'Rover' will not be the guilty party. The wise thing to do would be to look around before you punish him. To rush to judgment and execute could cost you far more than you might imagine. Even so, at the end of the day, your problem would remain unsolved.

Both the media and the general public have been quick to blame 'ozone' for all the 'perils of pollution.' But that is always a wrongful conviction. At worse, ozone could only be *guilty by association*. The true culprits usually lie elsewhere.

Myth #4 Ozone produces harmful toxins when added to water.

This should be an apparent fallacy just from the widespread and effective use of ozone in water treatment. But again, it is an exaggeration of the limited facts and therefore false association.

Bromate is considered the most important by-product of ozonation in water. It is formed from the oxidation of *trace* bromide ions in the water. Bromate has been shown to be carcinogenic in animals. Other by-products from ozonation may include aldehyde, bromoform and brominated acetic acids, none of which are classified as genotoxic carcinogens. However, and more importantly, studies have shown that at usual doses, only formaldehyde is produced in routinely measurable quantities and even that is formed at levels far below the WHO guideline concentration of 900 mg/lit.

Again, quite the contrary to this myth, a hallmark of ozone use for purification purposes, is the fact that it produces so little by-product. In fact, it is essentially an oxidizing agent, releasing the extra oxygen atom to leave a harmless oxygen molecule. Nothing could be more natural than that.

Myth #5 Ozone destroys the sense of smell.

What nonsense! Ozone is an effective deodorizer because, as an effective oxidizing agent, it eliminates many odors, especially the organic based odors such as are commonly found in the kitchen when cooking. The effect is not in the nose or the brain where the true sense of smell is experienced.

The effect is rather to chemically change the noxious chemicals by oxidizing molecules of carbon, hydrogen, sulfur and nitrogen, to odor-free molecules like carbon dioxide and water. Ozone does not mask the odor, it changes or neutralizes it. That's the ideal answer. But as we just observed above, you can, '*give a dog a bad name, and it can never do anything right*'. On the contrary, ozone does many useful things right! It is a key method for purifying the air - outdoors and indoors. And that we will see later, is an important strategy to obtain **Fresh Air for Life.**

Now we can move on to discuss the **Solution** in Part Two.

Part II
The Solution

Use these **FIVE ESSENTIAL STRATEGIES** to Purify Indoor Air

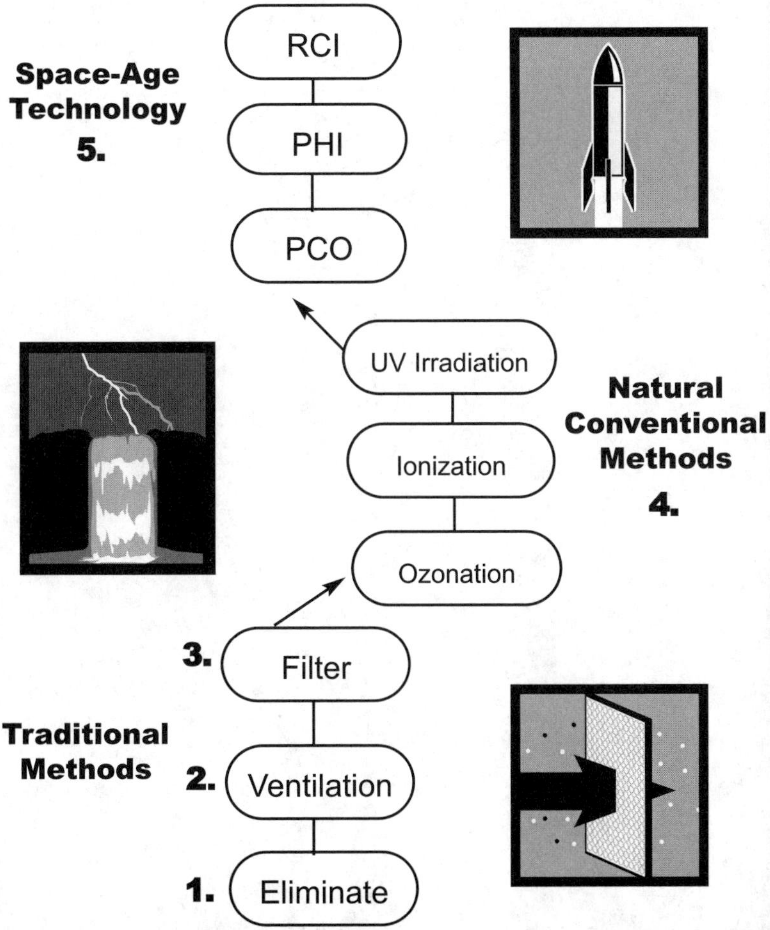

STRATEGY #1

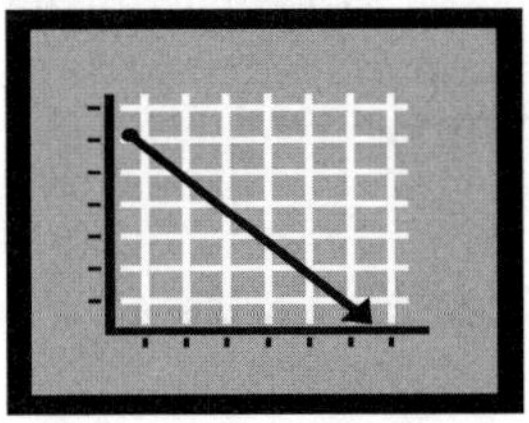

SIX

Eliminate Your Sources

*"Reducing Indoor Air Pollution is only **the first part** of the Solution"*

In Part One you would have discovered that there is a real Problem with indoor air pollution. The U.S. Environmental Protection Agency has identified this as one of the most important environmental health issues in the nation today. It raises serious *concerns* that you cannot afford to ignore, because *contaminants* you can't avoid will threaten you with health *consequences* that you cannot afford.

You have understood **the Problem**, so now we turn to address **the Solution.**

The U.S. EPA has also listed (1990) three main strategies for reducing indoor air pollutants: source control, space ventilation and air cleaning.(1)

Source control is considered, at least by some authorities, to be the most effective intervention overall. It aims to eliminate the sources of pollutants or at least reduce their emissions. Regrettably however, not all pollutant sources can be identified and practically eliminated or reduced. As we shall see shortly, they are too numerous, too diverse and too subtle at times.

Ventilation is effective because it brings outside air indoors. This is typically achieved by opening windows and doors, by turning on exhaust fans, or by using mechanical ventilation systems. We shall explore this strategy more fully in the following chapter.

Air cleaning is a general term but it must refer at least to doing something about the pollutants already in the air. *Particles and vapors* could be filtered out in several ways; *volatile organic chemicals* (VOC's) and other toxins could be neutralized physically or chemically, and *biological contaminants* could be neutralized or 'killed-off' in one way or another.

That's great, but when it comes to cleaning the air, we want to be more specific. There is new technology available today that could go much further than we could fifteen years ago. Therefore, we choose to be much more detailed and deliberate in our strategic plans. The general approach to the Solution of the indoor air quality problem that we will take in this book will consist of FIVE(5) ESSENTIAL STRATEGIES:

FIVE(5) ESSENTIAL STRATEGIES

1. **ELIMINATE the known Sources of Indoor Air Contaminants.**
2. **VENTILATE to dilute and exchange Polluted Air.**
3. **FILTER the air to trap Particulates.**
4. **PURIFY the air by copying Nature's Methods.**
5. **APPLY Space Age Technology to create practical Synergy.**

In the following chapters in Part Two of this book, we will explore these FIVE ESSTENTIAL STRATEGIES. You will gain an understanding of the development of an effective Solution leading up to our present state-of-the-art, space-age technology that provides real answers and offers hope for the future.

We begin in this chapter with an outline of some practical suggestions and ideas for reducing the major sources of indoor air contaminants that have been identified. Again, this is only the First of Five major strategies to finding an effective Solution to the Problem.

The leading candidate for elimination is probably the most obvious: **environmental tobacco smoke** (ETS). Much progress has already been made by public health promoters in increasing awareness and driving social action in this field. The intervention can be summarized in two simple and direct words: **Quit Smoking**! No single intervention in public health today could be more effective in reducing the promoting factors leading to primary disease prevention than the simple act of 'butting out'. No discussion about finding solutions to indoor air pollution would be complete without a fair exposé of this important lifestyle intervention. It is another 'sine qua non'. We must address this subject.

QUIT SMOKING!

The US Surgeon General has stated that 'smoking cessation' represents the single most important step that smokers can take to enhance the length and quality of their lives. There is no doubt about that.

If you smoke, you probably already know what you're up against. Health concerns usually top the list of reasons that people give for quitting smoking. Nearly everyone knows that smoking can and does cause lung cancer. But smoking also increases the risk of other major diseases, especially chronic obstructive pulmonary disease (COPD) and cardiovascular disease (CVD).

Based on data collected from 1995 to 1999, the US Centers for Disease Control (CDC) estimated that adult male smokers lost an average of 13.2 years of natural life and female smokers lost 14.5 years of natural life because of smoking. It accounts for up to 400,000 deaths each year in the United States.(2)

That's the bad news. Now here's the **good news**! No matter what your age or how long you've smoked, if you do smoke now, **quitting will help you to live longer**. People who stop smoking before age 35, avoid 90% of the health risks attributable to tobacco. Even those who quit later in life can significantly reduce their risk of dying prematurely. Ex-smokers also enjoy a higher quality of life with fewer illnesses and overall better health status. That far outweighs any risks from the average 5-pound weight gain or any adverse psychological effects (and withdrawal symptoms) that may follow quitting.

That news is only good for you if you now smoke and choose to act on it. A word to the wise should always be sufficient. But just in case, there's something more to be added in the context of indoor air pollution and its sources.

Smoking not only harms your health but the health of those around you. Exposure to **second hand smoke** (ETS or passive smoking) includes exhaled smoke as well as smoke from burning cigarettes. The U.S. Surgeon General has been warning of the deleterious consequences of involuntary smoking for two decades. Studies have shown that ETS causes thousands of deaths each year from lung cancer and heart disease in otherwise healthy non-smokers. Smoking by expectant mothers is linked to higher risk to their babies and after birth, smoke in the air leads to more sicknesses for young infants. Smokers who quit, definitely set a good example for their children.

Quitting smoking is not easy! But it can be done. Support from the society at large today is very favorable since smoking has definitely become less socially acceptable than it was in the past. You have more than adequate reasons to quit. There is a wide array of counseling services, self-help materials and effective medicines available today.(3) Smokers have more tools than ever before to help them quit successfully. **Therefore, you certainly can quit, if you want to!** You will do yourself a world of good. That simple though difficult change 'will add years to your life and life to your years'. Just as importantly, **you will reduce the pollution of the indoor air you share with others** who inhabit the same space. In so doing, you will enrich and enhance their quality of health and life. If after all, you won't do it for yourself, then for health's sake, *do it at least for those around you*!

FORGET THE FIREPLACE

North Americans have a love affair with fireplaces but it seems only in a romantic sense. That's why builders find it difficult to sell a new house without one and yet in 1997, only 2% of all households used wood as their primary fuel for space heating. The dream of a cozy fireplace to warm up to remains in dreamland. Most fireplaces are difficult to start and when they are fired up, they create smoke, sometimes produce unpleasant draft, cause unseen pollution problems and yet deliver heat to the home with only pitiful efficiency.(4)

Conventional fireplaces - just burning wood on an open grill beneath the chimney - are incompatible with new, tighter housing or with weatherized houses. They have large air requirements and produce toxic products of incomplete combustion. Beside the everlasting flame, smoke is forever coming off, usually from a part of the log remote from the flame itself. The incomplete combustion products become creosote in the chimney which can cause fires; they also turn into particulates and volatile organic products that can prove irritating if not toxic. Then there are the pollution gases: carbon monoxide, nitrogen oxide and dioxide, and oxides of sulfur. Add to all this, the real big dangers of:

- Increasing the *humidity* to favor dust mites and mold
- *Backdrafting* when the air supply or pressure is inadequate, allowing combustion by-products to flow back into the room, and in the extreme
- Potential house *fire* - a grave danger if there is creosote buildup or even worse, just plain carelessness.

Despite all this, field trials conducted by the Combustion and Carbonization Research Laboratory (CCRL) of fireplaces in Canadian homes, for example, proved that fireplaces actually had a negative energy efficiency in most cases.

That's enough information to make you forget the fireplace. But understandably, that's easier said than done. The deceptive urge to create ambience has compelled innovative solutions.

Some have retrofitted their fireplaces with 'tight' fitting *glass doors* along with a large combustion air supply directly from the outside to the 'firebox'. Then there is the artificial or manufactured *fire log* which lowers the high air demand, reduces pollutant emissions by up to 80% and lessens the chance of gas spillage into the house. But they are rather costly and provide more ambience than heat.

Woodstoves were a technical revolution. A well designed woodstove can fulfill most of a home's heating needs. They transfer heat primarily as 'black body radiators'. The early conventional airtight stoves averaged about 25 grams per hour of particulate emissions, about half that of the open fireplace. But this was still excessive and emission standards for particulates have now been set in both the U.S. and Canada. Nowadays advanced woodstoves give better combustion and have better heat outputs. They use preheated primary and secondary combustion air, along with well-insulated combustion zones. They have truly airtight, gasketed doors, special glass windows made from a pyro-ceramic material to transmit the infrared radiation from the flame to the room, and a hot air 'sweeping' of the window to allow clear viewing. That's all really high tech and they do deliver.

Gas burning fireplaces use solid ceramic logs placed among gas burners to produce that sentimental flame. Flue-gas condensation and chimney degradation tends to be high due to the high-moisture fuel, low burning rate and low temperatures. Draft can be a problem, causing combustion products to be brought back into the house. The best performers are the direct-vent fireplaces, with radiant transparent pyro-ceramic glass, good heat transfer to the house, an insulated outer casing and an effective venting system to ensure the safe removal of the combustion products.

If after considering all this, you are still enchanted by the romantic sight and aroma of that wood burning fireplace, especially as you come inside from the snow in the dead of winter, then at least take some sensible precautions:

- Have a trained professional inspect your fireplace at the beginning of every heating season.
- Always open flues when the fireplace is in use.
- If you choose a wood stove, make sure it is the right size and certified to meet EPA emission standards.
- Supplemental or 'decorative' heaters (including most unvented gas fireplaces) are not designed for continuous use. Never operate these appliances for more than 4 hours at a time.
- Use only seasoned hardwoods (e.g. elm, maple, oak) in fireplaces or stoves. They burn hotter and form less creosote. Avoid green or wet

woods and woods treated with preservatives.

- Make sure to supply adequate outdoor makeup air for combustion to avoid back-drafting.

Despite your best efforts, fireplaces almost inevitably contribute to indoor air pollution. This is one source that can be limited, if not simply eliminated altogether, by just a matter of choice.

Forget the fireplace!

MAINTAIN YOUR FURNACE

Whereas a fireplace in the modern home today may be little more than a romantic luxury, creating more ambience than actual heat, the same cannot be said of the typical furnace. In temperate climates at least, a source of indoor heat is an absolute necessity, especially when the freezing cold of winter settles down - not just for a night - but for a season. The typical home today in such climates is equipped usually with a central furnace that supplies heat to be distributed throughout the house. Both heating oil and gas are popular fuels today and both types of furnaces have many basics in common.

In both cases, fuel is mixed with air and ignited, heating a sealed chamber. Fresh, filtered air then blows across the outside of the hot chamber and is carried into heating ducts which then radiate the heat into adjacent rooms. In addition, and in some cases more importantly, the hot air is blown through registers in the floor (sometimes through the ceiling in basements or the walls when necessary) to bring heat directly into the room. The 'cold air' returns to the furnace usually through larger plenums inside walls. This is the common forced air system. The other (much older), system of radiators has boilers instead of furnaces and circulates hot water instead of air.

Furnaces require regular maintenance - a typical annual check up before the Fall season sets in, is both necessary and cost effective. It saves the system unnecessary wear, keeps it working efficiently and avoids the hazards of polluting the indoor air, especially with deadly carbon monoxide. Therefore, the annual check-up should be considered more vital to the family at home rather than to the furnace at risk. Your best bet is always to engage the services of a professional HVAC technician who will make sure all the bases are adequately covered. Since all the indoor air is essentially passing through the furnace, it is imperative that the potential for air contamination be minimized and safeguarded on a regular basis.

Let's consider a possible TEN STEP Furnace Maintenance routine, most if not all of it to be carried out by the professional technician annually: (5)

TEN STEP Furnace Maintenance

Step 1: *Shut down the system.* This is trivial but crucial. Accidents do occur and precaution is always a wise strategy in the face of change. Homeowners should always know the location of this switch in any case, in the event of fire or other emergency. Here it simply makes sense to turn the system off before performing any other routine.

Step 2: *Inspect the combustion air inlet.* Many furnaces draw air from inside the house to use for combustion. Therefore the furnace closet must be properly ventilated for adequate supply to the combustion chamber. Otherwise, the perils of soot and carbon monoxide can be realized. Homeowners are not unknown to block the inlets to reduce noise or heat loss inside the house. This is always dangerous folly. The inlet must always function adequately.

Step 3: *Clean the combustion chamber.* This is the chamber where the action takes place. Fuel mixes with the inlet air and the mixture is ignited to generate heat. But it also generates some soot, water vapor, carbon dioxide, and sometimes worse. Soot is scraped off with a wire brush and all loose debris is removed with an industrial shop vacuum. The chamber is inspected for holes or signs of corrosion.

Step 4: *Inspect the flue pipe.* The combustion products are directed through an exhaust vent to the outside. Holes and cracks are the enemy here. Small openings which would otherwise leak carbon monoxide and more, can sometimes be patched with foil tape. If ever in doubt, have them checked out. Corroded flues must be replaced. The damper is adjusted appropriately to moderate the draws of the exhaust up the chimney.

Step 5: *Replace the oil filter (where there is one).* This prevents tiny impurities from clogging up the oil-burning nozzle to create inefficiency and possibly shut down the furnace. This is simpler than it sounds. A valve is closed off and the replacement filter substituted easily.

Step 6: *Change the air filter.* Just imagine that all the air a family breathes in at home during the winter passes through the furnace filter. That's a big job except that the filter is adequate only for the furnace and never for the family. There will be much more about that later.

Nevertheless, you cannot change your furnace filter too often, at least once a year is mandatory. If not done, dust and dirt can work their way into the blower and coil assemblies, reducing efficiency and slowly ruining the motor. Apart from service maintenance calls, this typically accounts for half the calls of a furnace technician. This is a good time to also check the wear and tension on the belts that drive the blower to push the air through the duct system.

Step 7: *Adjust the burner and test efficiency*. This is probably the most technical step in the maintenance exercise and almost certainly demands some professional expertise. A combustion analyzer is used to examine the gases in the exhaust flue. This reflects the efficiency of the furnace itself and the fuel/air ratios can be optimized for maximum efficiency (and minimum pollution, by the way). In an oil system, the nozzle should be changed here. In a gas system, the burner tubes should be vacuumed clean.

Step 8: *Check the condensate drain.* The furnace produces water vapor which condenses into a drain pan which allows the water to drain out easily to contain the humidity level. If the drain is plugged, improperly attached or just sitting there, damp with debris, corrosion, accumulated dust or the like, then it becomes a breeding ground for mold and mildew. And you know what happens next. The spores and more are disseminated throughout the home by the blower and ductwork. And then … havoc!

Step 9: *Clean outdoor exhaust vents.* During the winter, snow and ice tend to accumulate from time to time around both intake and exhaust vents outdoors. If the vents become blocked for any reason, dangerous carbon monoxide fumes can back up into the house, with all that that implies. The furnace could also shut down. This step should be repeated often, especially when the winter is especially cold and snow generous.

Step 10: *Clean floor vents.* This is the only controversial step in the maintenance schedule we have just outlined. In principle, it seems like a good idea to remove floor registers and vacuum out the ducts. This should, in principle, remove dust, hair, food, scraps etc. from these sinks.

But not so fast. The question of duct cleaning remains somewhat in dispute and that's the next source of indoor air pollution that we shall address. Read on.

IGNORE THE DUCTS?

Regular cleaning of air ducts, in an attempt to improve air quality, has given rise to an industry that does make aggressive marketing appeal to concerned homemakers. But one must ask the fundamental question: Does duct cleaning improve air quality in homes and offices and help to prevent attendant health problems? The true answer is: No one really knows. There are examples of ducts that have become badly contaminated with a variety of materials that may pose significant risks to health. In such cases, the duct system can serve as a means to distribute these contaminants throughout a home. However, a light amount of household dust in your air ducts is normal.

The U.S. Environmental Protection Agency does not consider duct cleaning to be a necessary part of the recommended yearly maintenance of your heating and cooling system. That should consist of regular cleaning of drain pans and heating and cooling coils, regular filter changes and yearly inspection of heating equipment. **But there is little or no evidence to show that cleaning air ducts routinely improves air quality.**

The American Lung Association (ALA) takes a similar position. With increasing attention focused on health concerns from biological contaminants and dust in the indoor environment, the ALA offers the following recommendations regarding the use of duct cleaning:

"Duct cleaning has not been shown to prevent health problems, nor is scientific evidence currently available to conclusively demonstrate that particle (e.g., dust) levels in homes increase because of dirty air ducts ...

"When health problems are believed to be the result of biological contaminants or dust in indoor air, it is important to first determine that contaminated ducts are the cause of the health problems and verify that the ducts are, in fact, contaminated. The source of the problem may lie elsewhere, so cleaning ducts may not permanently solve the problem.

"People who have their ducts cleaned should verify that the service provider takes steps to protect individuals from exposure to dislodged pollutants and chemicals used during the cleaning process. This may involve using HEPA filtration when cleaning, providing respirators for workers and having occupants vacate during cleaning." (6)

If that's not conclusive, how about the scientific position of the homemakers' favorite magazine, *Better Homes and Gardens: ... 'knowledge about the potential benefits and possible problems of air-duct cleaning is limited because conditions in every home are different. The quality of air inside your home can be affected by cigarette smoking, cooking, and open windows. Even vacuuming the carpets can kick up household pollutants. With so many sources of indoor pollution, it's difficult to prove any role that ducts play in adding to the problem'.* (7)

C. P. Conwell, a forensic Industrial Hygienist has helpfully observed that air ducts get dirty because the particulates in the air stream drop out and remain in the duct. As the ducts become dirtier, turbulence in the duct increases and a greater proportion of particulates are retained. The duct acts like a filter and the junk remains stable there. Paradoxically, clean ducts are less efficient filters.

It would seem that cleaning air ducts essentially achieves two results. i.) It reduces the filtering efficiency of the duct system. The air coming out of the duct after cleaning is often dirtier than before cleaning. ii.) It distributes the material such that any remaining dirt really does have an increased possibility to re-enter the air stream.

In most properly operating systems, the relative humidity is not usually elevated enough to exclusively support mite or fungal growth. Therefore, it is uncommon for these organisms to colonize ducts, under normal circumstances. But if there is water loss, water damage or some type of water intrusion problem, that can complicate matters and should be addressed first. Otherwise, there is hardly any compelling reason to clean your duct system regularly. After a move or any major renovation or restoration, there may be some justification to clean the ducts while the house is unoccupied for a few days.

The question remains: Will cleaning your air duct system improve your air quality, improve your health and generally make good sense? Unlikely.

CLEAN YOUR CARPETS OFTEN

One of the consequences of the postwar industrial boom of the last century, was the development of widespread applications for synthetic oil-based resins and other polymeric materials. Among those developments, synthetic carpets spawned a major industry and so today, carpets are spread throughout homes and commercial buildings everywhere. Wall-to-wall broadloom adds a touch of class and elegance to almost any room and the variety of style, color and texture is a decorator's dream. But with this modern floor covering has come a source of indoor air pollution that includes:

- Odor and off-gassing of organic vapors from the carpet backing and adhesive
- Retention of allergens from dust mites and pet dander, especially with increased humidity
- Fungal contamination from poor maintenance or water damage

Most carpets today (contrasted with traditional woolen rugs) are made of synthetic pile, usually nylon or polyolefin (e.g. polypropylene), cut pile or nylon tufted through a primary backing with a back coating of solvent-based adhesive. Carpet tiles are usually made with a hard backing of PVC (polyvinyl chloride) or a hydrocarbon resin.

New carpets tend to emit volatile organic vapors from the solvents used in their manufacture, as well as the latex backing and the glues used to hold them down. The typical 'new' odor is that of 4-phenylcyclohexene(4-PCH) which is a by-product of the styrene/ butadiene latex binder. This volatile chemical is detectable at very low (trace) levels by the human nose. The odor can be quite uncomfortable and annoying, so fortunately, by avoidance, 4-PCH has not proven to be a real health hazard at these low levels. It really dissipates within a week after installation and should be undetectable after 2-4 weeks.

Carpets are indeed, a primary source of indoor air pollution

On a regular basis, carpets (and other textiles like upholstery, bedding and drapes) act like a "sink", collecting airborne pollutants and aerosols. These 'fleecy materials' absorb and later re-release different kinds of contaminants: allergens, chemicals and mold spores. If they are chronically exposed to dirt and moisture, they tend to be a haven for mold and dust mites. The same thing happens if the sub-floor has excessive moisture. Dust mites love high humidity environments anywhere indoors, and carpets are no exception. However, in most residential environments, bedding provides the primary source of dust mite exposure. But whenever a carpet remains damp, the other headache to deal with is the growth of mold - sometimes toxic mold. As such, carpets are indeed, a primary source of indoor air pollution in most homes.

The national trade association for the carpet and rug industry - known as the *Carpet and Rug Institute* (CRI) - has a voluntary indoor air quality testing and labeling program that tests chemical emissions of new carpet. Carpets that satisfy the standard limits for emission carry a special "CRI" icon and should be the first choice of consumers.

Here are TEN PRACTICAL TIPS to reduce the level of indoor air pollutants attributable to carpets. They are part of the total solution.

TEN PRACTICAL TIPS

to reduce the level of indoor air pollutants attributable to carpets

- **If you choose to use carpets** on floors, insist on CRI-certified products only.
- Air out the area thoroughly and remove all evidence of moisture anywhere, before installation of new carpet.
- Insist on carpet tacks (rather than glue) wherever possible.
- Vacate the area before, during and up to 7 days after installation.
- After installation, arrange for maximum ventilation with fans and air conditioners for at least 2 days following installation.
- Protect carpet from all potentially hazardous products (use plastic covers, walk-off mats and remove shoes whenever necessary).
- During construction or remodeling, install the carpet as the last possible task on the schedule.
- Maintain humidity levels between 40 and 60% to decrease dust mites and mold growth.
- Vacuum carpets frequently with a high efficiency filtration system (HEPA) or better still, use a central vacuum system that extracts to the outdoors.
- Use only approved carpet cleaning systems that are verified to leave the carpet dry no later than 12-24 hours after cleaning, and one that will remove more than 98% of dust, surface mold and known (mite and pet) allergens.

Better yet than all the above, you could consider using hardwood, ceramic or other hard, washable floor surfaces? That too, can certainly be designed to have a charm of its own … a healthy charm indeed!

THE PET LOVER'S DILEMMA

Here we are again. Pet owners have this love affair with their pets. Even when their pet allergies or asthma cause them to suffer, they choose the latter rather than get rid of their pets, more often than not. However, as much as you may love your pet, chances are that it will contribute to indoor air quality problems.

Pets shed dander or skin flakes, which contain proteins that cause allergies. Their saliva, urine and feces also have protein allergens. Their hair does not usually, but hair or fur can collect pollen, dust, mold and other allergens too. Feathers and droppings from birds also increase allergen exposure.

In light of all this, people who suffer with environmental allergies or asthma should be cautious about what type of pet they choose to bring into their home. The best type of pet for the allergic patient is probably tropical fish but then, large aquariums could increase humidity and favor the growth of molds and dust mites. So, you can't win.

Nevertheless, **more than 70% of U.S. households have a dog or cat.** So what can be done to improve the home environment? Where pets are known to cause significant allergic reactions, they should be removed to avoid progression of symptoms. If that option is followed, a thorough cleaning of floors, walls, carpets and upholstered furniture is very important. Despite best efforts, it still takes an average of 20 weeks for allergen levels to reach the levels found in homes without pets. A "trial removal" of a pet for a few days or weeks would really be of little value.

If your family is unwilling to remove the cherished pet, it should be kept preferably outdoors, or at least out of the bedrooms. Isolation is never adequate however, since pet allergens circulate in the air and pervade the whole house and the inhabitants' clothing.

What else can be done to reduce pet allergens? Here are **TEN PET SUGGESTIONS:**

TEN PET SUGGESTIONS

- If you are allergic, do not pet, hug or kiss your pet, for obvious reasons.
- Set and enforce boundaries for your pet's range and activities.
- Keep pets away from fabric covered furniture, carpets (where possible) and stuffed toys.
- Vacuum carpets, rugs and furniture two or more times per week.
- Use a High Efficiency Particulate Arresting (HEPA) filter on the HVAC system and vent to the outdoors if at all possible. This is often not an effective solution, especially if individuals have severe symptoms.
- Wash your pet frequently. Allergic individuals could have some-one else do this for them.
- Place litter boxes in an area unconnected to the home's air supply. Clean frequently.
- Avoid masking odors with more airborne chemicals.
- Take appropriate precautions when visiting other people's homes.
- Use a proven air purifier with space age technology. You'll need one.

LIMIT YOUR USE OF CHEMICALS

The household cleaning agents, personal care products, pesticides, paints, hobby products and solvents that are so convenient and make our lives so easy are also sources of potentially harmful chemicals that pollute our environment. The range of these that contain potentially harmful substances which contribute to indoor air pollution is wide-reaching and diverse. And so are the health consequences. They arise from the combined effects and with prolonged exposure.

Here's a simple exercise. Do a walk-through of your home in search of all the household and personal care products that contain volatile solvents, aerosols and other organic chemicals that can prove toxic to the home environment. Go systematically through the kitchen, the bathrooms, the storage cupboards, the basement and finally the garage, as if you were a private investigator on a mission. You'll be amazed at what you would find. Here's a short list of five important classes of indoor chemical pollutants and some implications.

Household Cleaners. The average American uses about 25 gallons of toxic or hazardous chemical products per year in the home. Most of these can be found as common household cleaning products. As a nation, we pour some 32 million pounds of household cleaners down the drain each day. Whether we use them for cleaning dishes, laundry, bathrooms, ovens, other appliances, floors, windows, barbecues or what have you - just having these conventional products existing in the house environment is a 'toxic' accident waiting to happen. Here's a brief list of some common ingredients that are known to be hazardous:

Ammonia	Lye	Propylene glycol
Bleach	Naphtha	Sodium Hypochlorite
Chlorine	Nitrobenzene	Sodium laurel sulfate
Formaldehyde	Perchloroethylene	Sodium tripolyphosphate
Hydrochloric acid	Petroleum distillates	Trichloroethane

Some of these products have been linked to a vast array of catastrophic sicknesses and disorders including:

Altered metabolism	Hormonal changes
Enzyme dysfunction	Birth Defects
Nutritional deficiencies	Developmental disorders
Learning and behavioral disorders	Cancer

A report by the UCLA Pollution Prevention Education Research Center identified adverse health effects for a number of common chemical ingredients found in some 222 different cleaning products. The environmental and occupational hazards associated with many of these chemicals are not well understood but the Center's Director raises "increasing concern because they are *sources of indoor air pollution* and multiple occupational exposures, and ... may be released ... into the ambient environment." (8) Inhalation is a major route of exposure for many of these toxic chemicals in everyday use.

A joint project between the U.S. EPA, California and local officials found that 41% of standard cleaning chemicals are 'dangerous' and some are 'too dangerous to use'. EPA researchers have called on manufacturers to use less lethal ingredients and improve the safety of packaging for products including those designed for bathroom and kitchen use.

Household cleaners are the most frequent chemicals involved in poisonings reported to 'poison control centers' in the United States. The U.S. EPA has reported that toxic chemicals in these household cleaners (many of which create exposure to fumes) are three times more likely to cause cancer than outdoor air pollution. It's no wonder that a 15 year study reported at the Toronto Indoor Air Conference (1990) that women who work at home have a 54% higher death rate from cancer than those who work away from home.(9)

So, what can you do in the face of such a threat? Surely, you can begin to take control of the situation. The following list of ideas is just the obvious starting point:

- Look for safer, cleaning product alternatives.
- Use chemical cleaners only when absolutely necessary.
- Read all labels carefully before using. Be aware!
- Leave all products in their original containers with labels.
- Do not mix cleaners.
- Avoid them in pregnancy as much as possible.
- Use only in well-ventilated areas.
- Do not sit, drink or smoke while using these products.
- Thoroughly wash and rinse, and clean up after use.

Air Fresheners and Deodorizers exploit chemical properties to mask unpleasant odors. They generally *either* hide one scent with another more pleasant fragrance or else they simply desensitize the nose. They do absolutely nothing to improve actual air quality. But at the same time, they are not harmless. Their chemical ingredients can be irritating and can trigger allergies and asthma attacks. Common p-dichlorobenzene is known to cause cancer in animals.

What can you do? Here's another list of practical suggestions to get you started:

- Avoid the use of these products.
- As an alternative program, try addressing the cause of any unpleasant odors.
- Simmer cinnamon and cloves in the kitchen and scatter natural potpourri in open dishes elsewhere.
- Sprinkle baking soda in the cat's litter box and do the same on carpets the night before vacuuming.

- Place an open box of baking soda in the refrigerator and in closets and bathrooms.
- Sprinkle borox in the bottom of garbage cans to inhibit molds and bacterial growth.
- Finally, you might insert used lemons with the garbage.

Personal Care Products are just that: intensely personal. Responses to them tend to be individualized. One person's delight and fancy is another person's worst nightmare. These products use thousands of different chemicals and some single products can actually contain hundreds. Even those products advertised as 'gentle' and 'soothing' like *baby shampoos* contain some of the worst chemicals. For example, a baby shampoo might contain anaesthetizing agents to cover up the burning sensation that would otherwise be caused when the chemicals contact the eyes.

But every *chemical* solution poses a new chemical problem. Volatile solvents like acetone, alcohols, formaldehyde and benzene, are very common and the use of aerosol sprays only makes matters worse. Many of the chemicals in these products are just downright dangerous and as a class, they should be minimized if not avoided altogether. With good hygiene practice, unpleasant body odor is not as prevalent a condition as one might think. So this tends to be *vanity* at work, more often than *necessity*.

To help reduce this source of indoor air pollution, consider doing your part, as the rest of us all do a little here and there, to minimize the problem. After all, in the workplace this can be very counter productive. But for health's sake, let's all do better, by choosing the more natural alternatives most of the time. *Caution*: if you work in this industry, your risk of exposure goes up dramatically. Therefore, be alert and take defensive action.

Laundry products also contribute to poor indoor air quality. Detergents contain phosphorus, enzymes, ammonia, naphthalene, phenol, and many other chemicals. The typical residue left on your clothes, linens, towels etc. is absorbed through the skin and often causes allergic dermatitis (skin itch and rash). Some of these chemicals also become airborne and are inhaled. Here are some practical suggestions to reduce this source of contaminants:

- Use liquid detergent that do not contain phosphates. Add borax, washing soda or even baking soda for a better wash.
- Use non-chlorine dry bleach or hydrogen peroxide-based liquid bleaches.
- Use fabric softener sheets preferably, they release less toxic chemicals into the air than liquids or aerosols.

- Avoid chemical stain removers.
- Don't hang laundry indoors - Humidity favors mold growth.
- Vent appliances to the outside and maintain gas appliances.
- Check out LaundryPure® - a revolutionary space-age, environmentally friendly alternative to doing laundry, without detergent.

Pesticides pose a special challenge to more than 90% of U.S. households, especially for those with small children. They are the #2 cause of death by poisoning in the United States. Inhalation can have major immediate symptoms. These chemicals are stored in body fat and can damage organ systems in the long term or even become fatal. They come into the home from outdoors, floating in as contaminated dust or tracked in on shoes or clothing. Otherwise, they are stored in-house, or used and persist as volatile residues. Whether they are sold as sprays, liquids, sticks, powders, crystals, balls or foggers, the risk of exposure remains. They should be treated with great caution, as dangerous toxins to be avoided as much as possible. Here again are a few suggestions as to what you can do to reduce exposure:

- Resort to pesticide use only when absolutely necessary.
- Use professionals where possible, but screen them.
- Use alternatives to chemical pesticides wherever possible.
- Do not store pesticides inside the house.
- Close windows and doors during and after application outside.
- Watch out for cross-contaminants on shoes, gloves or clothing.

There you have it. Household cleaners, air fresheners and deodorizers, personal care products, laundry products, pesticides - that's just a short list. American homes are typically crowded with many more chemicals that are stored and occasionally used for more purposes than we care to detail here.

Reducing the sources of indoor air contaminants in every form is clearly a First Step along the way to finding an effective Solution. It is but **Strategy #1** in the **FIVE ESSENTIAL STRATEGIES** en route to real **Fresh Air for Life!**

Indoor air pollution is real and despite all attempts to reduce the known sources in homes, offices and institutions, the challenge obviously remains. That takes us then to the Second major approach (**Strategy #2**) in dealing with the Problem: that is, the need to **Improve Ventilation** - the way to create adequate exchange that allows *fresh air* in and *foul air* out.

That's the subject of the next chapter.

STRATEGY #2

SEVEN

Ventilate Your Space

"Replacing indoor contaminants with outdoor pollutants ***cannot be the final solution."***

Any attempt at Reducing the Sources of contaminants like we have described in the previous chapter, is just the *First* of **FIVE ESSENTIAL STRATEGIES** that we are proposing to combat the hazards and consequences of indoor air pollution. Now we turn in this chapter, to address the *Second* of these major strategies: **Improving Ventilation.**

But immediately, we should define some common terms.

Ventilation is defined as 'the process of supplying and removing air by natural or mechanical means, to and from any space.' Sometimes the word is used to describe only the outdoor air provided to a space. That is correctly, 'outdoor air ventilation". More generally 'supply air' or 'total air' ventilation includes both the outside air *and* 're-circulated air'. The *ventilation air* (a volume of air in unit time) is given in liters per second (L/sec) or cubic feet per minute (cfm), to quantify the amount of ventilation in a given space.

Natural ventilation refers to ventilation through intentionally provided openings, such as open windows, doors and vents. The flow of outdoor air that comes through unintentional openings is referred to as *infiltration*. This can take place through cracks around window frames and wall-to-floor joints, and so on. *Mechanical ventilation* is ventilation provided typically by circulating fans, with or without a duct system. It usually provides a supply of air from the outdoors and/or an exhaust system to the outdoors.

Indoor air quality problems can often be traced to a simple condition: *Lack of Fresh Air*. Inadequate ventilation can make indoor conditions much more dangerous. Whether you think of natural or mechanical ventilation, there are essentially **three reasons** why any building or enclosure might need to be ventilated. The first is the provision of outdoor air ***for health and comfort*** of the occupants. We need air and we need Fresh Air, all the time. Stagnant, re-circulated air is both unhealthy and uncomfortable. A breath of Fresh Air is always refreshing and also necessary. So, think Fresh Air for Life!

Secondly, ventilation is one method to deal with polluted indoor air. It permits ***removal of internally generated contaminants***. Those contaminants pose a real Problem as we have already seen in Part One of this book, and the potentially serious consequences must be avoided at all cost. Thirdly, ventilation maintains ***specific pressure relationships*** between certain indoor spaces and between these spaces and outdoors. Failure to do this adequately can lead to dangerous in-drafting, for example, that could create serious problems, as air in the indoor spaces become really hazardous and unbreathable.

It might seem obvious and even simplistic that if the indoor air is polluted, the thing to do would be to open all windows and doors to let in some fresh air. This should dilute the polluted air at least, and at best, replace it with fresh air. Any means to facilitate the movement of air, such as by a mechanical fan, should only enhance the speed and efficiency of this ventilation process.

But once again, things are not that simple. In fact, the development of understanding regarding ventilation and the application of this technology was a very long time in coming. You might be interested to learn just a bit about this history.

A BRIEF HISTORY OF VENTILATION[1,2]

After fire was invented and eventually brought indoors, the problem of indoor 'smoke' arose. Ventilation, in a sense, was discovered as a solution to that problem. An opening in the roof let out the smoke and allowed air to come in to keep the fire burning. However, believe it or not, formal chimneys did not come about until the twelfth century.

The ancient Egyptians recognized that stone carvers who worked indoors suffered more respiratory distress than those working outside. They concluded that this was due to the dust inhalation and practiced primitive ventilation to reduce the 'occupational hazard'.

The Greeks, as usual, were a little more philosophical about it. They distinguished 'good' air from 'foul' air but despite debate, could not agree whether 'good' air was really indoor or outdoor. Hippocrates postulated that decaying organic matter from marshes and wetlands created an infected air or '*miasma*' that was the carrier of pestilence. Up to the fifteenth century, 'foul air' was considered as 'this natural condition to be avoided by proper city planning and limited building away from marshes'.

But with urbanization of Western Europe after the Renaissance, the contaminants and fumes from fossil fuel combustion and toxic manufacturing processes produced a new 'fuliginous vapor' and 'horrid smoke' that had to be included in the definition of 'foul air'. With that came offensive and irritating odors, as well as a number of chronic and epidemic diseases. People began to realize that air in a building could *somehow* transmit disease among people in crowded rooms. This led to the deliberate sealing of 'sick rooms' and the popular avoidance of dangerous and polluted 'night air' which was believed to cause tuberculosis.

Smoke, dust, disease - those were the obvious problems. King Charles I of England, for example, decreed in 1600 that no building should be built with a ceiling height of less than 10 feet (or 3 meters) and that windows had to be higher than they were wide. All this was an attempt to improve smoke removal. But there remained no understanding. When charcoal was found to burn with little smoke, it became the preferred fuel of the wealthy. In their presumptuous ignorance, they burned charcoal in braziers placed in closed rooms without ventilation and caused people to die in their sleep from carbon monoxide poisoning. This included, by some accounts, Philip III of Spain.

Fireplace design did not improve much until the end of the eighteenth century when Benjamin Franklin used his Pennsylvania Fireplace - the first closed stove to effectively heat an unsealed room, and another Benjamin (Thompson, this time, but actually Count Rumford) introduced the smoke shelf to the open fireplace. With these developments and others, by the end of the eighteenth century, organic contamination of indoor air had come to overshadow the issue of smoke from combustion.

During the late eighteenth century, another major advance took place. Joseph Priestly isolated the oxygen in air for the first time and another chemist, Antoine Laurent Lavoisier discovered carbon dioxide. That ignited a debate which lasted almost a hundred years. The central question before the house was this: Is *'bad air' caused by oxygen depletion or excess carbon dioxide?*" Many years later, Pettenhofer concluded in 1862 that it was neither oxygen nor carbon dioxide that was responsible for 'bad air'. It was biological contamination.

It is useful to note that in any case, it was human contamination of the indoor environment that was the problem. As we said in Chapter I, '*we have seen the enemy long ago, and the enemy is us*'. It is our breath, our bugs, our bad habits ... that pollute the air. **We trap ourselves indoors and thus wage war with ourselves in the arena of our immediate environment, the surrounding air**.

Now, guess what? You thought the Greeks were argumentative. They were not alone. That new emphasis on biological contamination re-ignited another debate that lasted for several centuries. It pitted the architects and engineers on one side, against physicians on the other. The first group were obsessed with *providing comfort and freedom from noxious odors* and the debilitating effects of oxygen depletion and/or carbon dioxide accumulation. That's all well and good. But physicians were (correctly, I dare say) more concerned about *minimizing the spread of disease*. They wanted more ventilation!

Ventilation rates are measured in cubic feet per minute or cfm and guidelines are usually described in **cfm per person**. Dr. I. Billings, the physicians' lead influence in those early days, argued for a standard of 60 cfm (28L/sec) of ventilation air per person in his concern for disease prevention. That was twice as much ventilation as required for the normal comfort level of 30 cfm (14L/sec).

By the beginning of the 20th century, understanding was improving and people began to see the light. Two other great causes of discomfort in the indoor environment were added to the mix: excessive temperature and unpleasant odors. By 1925, 22 states in the American Union had stipulated a minimum ventilation standard of 30 cfm (14L/sec). Resistance to higher levels of ventilation came from the energy conservationists who objected to heating large quantities of cool outdoor air for ventilation in the northern climates.

Proper ventilation required engineering solutions. Opening and closing windows and doors or building vents and shutters was altogether, simply inadequate. First designs took advantage of natural ventilation induced by heat sources. The classic example of this was the innovation by the famous architect, Christopher Wren, who designed new *holes in the ceiling* of the British Houses of Parliament to effectively relieve the heat generated from candles in 1660. Then came *aspirating ducts*, followed by *shafts* for heating and ventilating, then a combined shaft for both. Add *steam coils*, and mechanically driven *fans* and put it together in a '*blower system*' and you had the genesis of a modern HVAC system. Cap it off with full *air conditioning* in the 1950's and you're into the modern era.

Actually, the first air conditioning system dates back to 1902 when Willis Carrier designed one for a printing firm in New York. By the 1910's, the technology moved into movie theatres which became the first public places to enjoy this luxury. A decade later, air conditioning systems were being fitted into department stores. Single unit air conditioners did not come into any kind of regular domestic use until just before World War II. In fact, by the mid 1960's only ten percent of American homes were air conditioned. By 1995, that number was up to 75 percent. These are only modern conveniences.

The modern HVAC system is an integrated one, combining the processes of heating, ventilation, air conditioning and (usually) humidification in one central system. The technology is much more sophisticated, especially for the furnaces and air conditioning units, and their computerized solid-state controls. But the essential operational concepts have remained more or less unchanged. Air is drawn in from outside, pre-conditioned and then heated or cooled, as required. It then circulates via ducts and registers, through the occupied living space and is then returned for further filtration and reconditioning, while some is vented to the outside. The central system allows for other air treatment methods to be incorporated, such as filters, humidifiers and electronic air cleaners. These are almost standard amenities in the modern American house.

It is our breath,
our bugs,
our bad habits ...
that pollute the air.
We trap ourselves indoors
and thus wage war
with ourselves in the
arena of our
immediate environment -
the surrounding air.

In the first half of the twentieth century, building design and construction also changed to the more modern styling. Reusable masonry products were steadily replaced with impenetrable curtain wall buildings using unprecedented materials like glass, steel, plastics and lighter masonry and concrete panels.

But the real kicker came in the 1970's with the famous oil embargo that led to an energy crisis. Energy was in short supply and that changed everything. Certainly it changed the attitude to environmental control. Now it cost a lot more to heat a building, so bringing in outside air in cool weather (or the reverse in hot weather, with the demand for air conditioning) was an expensive proposition. Add to that the cost of distributing the incoming air and it had a '*double whammy*' effect.

The impact was so serious that it prompted even the American Society of Heating, Refrigerating and Air Conditioning Engineers (ASHRAE), within a few months, to rewrite a new standard for ventilation, down to 5 cfm per person, across the board for new buildings, in non-smoking environments (but they raised it to 25 cfm in smoking environments). They revised the standard yet again in 1989 to 15 cfm for both smoking and non-smoking environments. It is now qualified by the size of the building space and its functional type. In addition, the emphasis is on control of energy distribution within buildings - shifting from the approach of constant air volume to a variable air volume strategy. But that's beyond the scope of our discussion here.

Some studies estimate that approximately 60% of all indoor air problems are due to inadequate ventilation. Of this 60%, 30% are a result of air contamination - 20% from inside the building and 10% from outside.(3) Another study by NIOSH attributed half the cases of indoor air pollution to poor ventilation.[4] The U.S. Government Accounting Office estimated 20% of U.S. schools have indoor air problems while 25% have unsatisfactory ventilation.[5]

We're interested in understanding the principles of ventilation only to deal effectively with the problem at hand. That is, if indoor air is polluted - and we know it is - what can we do to dilute or exchange it with fresh air? We therefore turn first to the basics of natural ventilation.[6]

NATURAL VENTILATION

Many buildings throughout the world are naturally ventilated. Openings created by design and construction allow for the natural flow of air into and out of the building. In many cases, the net results are poorly controlled and unreliable.

But we do have a good understanding of basic principles that allows for efficient design and control with some predictable reliability. We do understand the driving forces behind the movement of air, plus the impact of different types of openings and that allows us to devise effective techniques for maximum ventilation. The results will naturally fluctuate in time.

Driving Forces

The driving forces of natural ventilation are wind or ***wind pressure*** and the difference between indoor and outdoor temperatures. In simple terms, when wind strikes a building, it induces a positive pressure on the front or windward face and a negative pressure on the back or opposite face. There is negative pressure also on the 'wake' region of the side faces. The net result is that it causes air to enter openings on the front, pass through the building from the high pressure windward areas to the low pressure downwind areas. In other words, **wind moves air from outside, through the building and back out the building on the opposite side.**

Stack pressure is quite different. The effect results from differences in air temperatures and therefore air density between the inside and outside of a building. This causes an imbalance in the vertical pressure gradients of the internal and external air masses. The way it works is this - If the inside air temperature is greater than the outside, air enters through the lower part of the building and escapes through higher openings. When the air temperature inside is lower than outside, the flow direction is reversed.

In practice, systems are designed to allow both of these effects of wind and stack pressure to work together complementing each other. In addition, to obtain efficient natural ventilation, buildings are designed to be otherwise airtight (with minimal infiltration) so that ventilation can be confined to the flow of air through 'intentional openings' only. As the professionals usually say it, when designing a ventilation system, one wants to **"build tight and ventilate right".**

Speaking of professionals, you might be interested to know that natural ventilation systems can be modeled. One useful approach is to treat the indoor space as an air quality reservoir. As such, any transient diffusion of indoor air pollution can be initially accommodated by diffusion (and therefore dilution) through the enclosed air mass itself. This can then compensate for fluctuating ventilation over time, under certain circumstances. This becomes important particularly with variable weather conditions on the outside. Just imagine the changes when the temperature drops or the wind ceases. The weather is no more certain than it is predictable.

Openings

With natural ventilation, the idea is to have the outside air enter through intentional openings, designed for maximum efficiency. A good design would have two kinds of vents. Permanently open vents would provide background ventilation and then there would be controllable or adjustable openings to satisfy any transient demands. These latter openings could be adjusted by automatic controls or with dampers.

In many buildings around the world, the principal component of natural ventilation is simply windows that can be opened or closed. In the tropics and during summer months in temperate climates, the windows allow large air masses to flow through to purge and cool the building naturally. But when winter comes, these same windows waste energy and sometimes cause discomfort if they can't be controlled and sealed properly. If the outdoor air is itself polluted especially in urban areas, near busy traffic or in the vicinity of industrial sites, **these windows can also be a source of indoor pollution itself, coming in from the outside.** Smoke, exhaust, ash, VOC's, odors and even microbes are potential invaders through these welcoming apertures. Then it becomes bad news.

Some air vents or 'trickle' ventilators are preferred options in place of window openings, for *winter* ventilation especially These are typically permanent openings about 4000-8000 mm^2 in size and with or without manual adjusters. At least one vent per room is usually recommended for naturally ventilated buildings. In Britain, it is recommended that office buildings have 4000 mm^2 of opening for each $10m^2$ of floor space. The ventilators in this case are placed at a higher level and positioned for good air mixing and distribution, and sometimes over space heaters conveniently.

Other air inlets can be automated in response to different air quality or climate parameters. In other words, areas of opening can be automatically adjusted in response to sensors of temperature, humidity or pressure. The vents are adjusted if the outside temperature falls too low, or the moisture in the room is too much, or the pressure difference is too high. All this is implemented in order to maintain a fairly uniform flow rate and a comfortable and healthy indoor environment.

Where combustion (especially open combustion) takes place in a room, regulations demand a minimum area of permanently open vents (with or without flaps) in order to guarantee an adequate supply of air for combustion and efficient exhaust. This is an essential safety measure to protect against back draft or suction pressure but at the same time, it is an additional energy cost due to increased air exchange. Room-sealed combustion appliances with balanced flues or externally supplied and exhausted air make for better efficiency, less pollution and good common sense.

Design

The physical arrangement of openings is a matter of engineering design and beyond our interest here. But suffice it to point out that generally more than one opening is required for good natural ventilation. Adequate spacing is required to generate reliable air exchange. Some designs, especially in moderate to medium cold climates exploit passive stack ventilation. Air flow is driven through the stack by a combination of stack pressure (as we described earlier) and wind-induced suction pressure. In some countries, where prevailing wind provides a reliable driving force, a stack may be arranged as a wind tower. The openings face strong incoming winds so that the wind-driven airflow is actually ducted into the building.

Let's put all this together. Since local climate conditions are variable, natural ventilation solutions should be designed to be robust and capable of satisfying indoor air quality needs and making the space environment most comfortable. This calls for some minimum and maximum requirements. The former assures the better indoor air quality while the latter provides the adequate summer cooling. In the ideal case, the minimum requirement would be satisfied with permanent openings, while the maximum need would be met with adjustable openings.

Table 5 summarizes the major advantages and disadvantages of Natural Ventilation.

Table 5. Advantages & Disadvantages of Natural Ventilation

Advantages

- Suitable for many types of buildings in mild or moderate climates
- Open windows popular in pleasant locations and mild climates
- Usually inexpensive
- High air flow rates possible
- No need for plant space
- Minimal maintenance
- Short periods of discomfort are tolerable

Disadvantages

- Inadequate control
- Difficult with deeper multi-roomed buildings
- Unsuited to noisy and polluted locations
- Security risk
- No filtration or cleaning of incoming air
- Impractical in severe climatic regions
- No heat (energy) recovery

MECHANICAL VENTILATION

Natural ventilation has distinct advantages in some locations or situation, but in others, one is obliged to resort to mechanical ventilation.(6) **The big advantage here is the element of control.** These systems can be designed or operated to mix or dilute pollutants (the *mixing* mode) or else to displace or remove pollutants without mixing (the *displacement* mode). In any case, the mechanical ventilation system could be made to be almost independent of varying climate conditions. **But they cost more** and despite good maintenance, they eventually have to be replaced.

The key ingredients of the mechanical ventilation system are fans, ducts, air intakes and diffusers, plus or minus silencers or noise alternators where necessary.

Fans drive the system. For systems of low capacity, propeller fans are most common, whereas for high capacity and lengthy ducts, the centrifugal and axial fans predominate. In either case, they consume a significant amount of energy. It turns out that with wider ducts, the power consumption of fans is much reduced, but those ducts can add to space and cost requirements. The ducts that transfer the air provide a resistance to flow and have their own additional requirements for design, sealing and insulation. They connect to air intakes that collect the outdoor air. Needless to say, those intakes should be away from pollution sources like traffic fumes, industrial output or other building exhausts.

Mechanical ventilation can be a source of noise pollution and attempts at sound proofing the ducts or using silencers can prove very helpful.

Two basic configurations are used in mechanical ventilation. One focuses on removing the indoor air by forced extraction, whereas the other addresses the supply or introduction of outside air, again by mechanical means.

Extract ventilation

A fan can be used to suction indoor air which in turn induces an under pressure that promotes the flow of an equal mass of outdoor air -hopefully fresh air - to replace it. For best efficiency and control, the goal would be to keep the mechanical pressure just above the weather-induced pressure. If the building is airtight, then only the appointed air inlets supply the make up air and it is possible, in principle only, to heat, cool or filter the incoming air as necessary. But this is seldom the case.

Mechanical ventilation is often used to extract air in local situations where pollutants or excess moisture are being produced. It is not uncommon either to use local extractors to supplement natural ventilation systems - especially for domestic use. They are common above cooker ranges in kitchens, or in low-capacity walls and windows in bathrooms, attics and sometimes basements. On the other hand, the corresponding air inlets tend to be located in living rooms and bedrooms. They can be operated intermittently as required, using a humidity sensor or time switch for example, as automatic control. Needless to say, mechanical ventilators are also important to avoid the spread of serious local contamination as in laboratories, hospitals, or in industrial areas with volatile chemicals.

Table 6 summarizes some major advantages and disadvantages of Mechanical Extract Ventilation.

Table 6. Advantages & Disadvantages of Mechanical Extract Ventilation

Advantages	Disadvantages
• Good control	• Costs (system, operation)
• Pollutants removed at or near source	• Noise could be annoying
• Reduces moisture effectively	• Risk of back draught
• Possible heat recovery	• Risk of indrawing radon and soil gases

Supply Ventilation

In contrast to extract ventilation where air is removed by force, the alternative is to mechanically supply or add outdoor air into the building where it mixes with existing air. It's a '*push*' mechanism compared to the previous '*pull*'. This increases inside pressure which displaces indoor air through openings that may or may not be intentionally provided. The system can be designed in such a way that the incoming air can be conveniently pre-cleaned and thermally conditioned.

Most supply systems are provided with ducts and the air is filtered to reduce dust and particle contamination. For effective control, the same air tightness and vent conditions are needed as for extract ventilation and the system is again best operated at pressures just above the weather-induced pressures. It is particularly valuable as a strategic design when the outdoor air is very polluted as in urban or industrial settings. The incoming air is then appropriately 'treated'. It is also useful for preserving clean rooms in industry and healthcare facilities as well as in situations where occupants of a building suffer greatly from allergies. Of course, that has to be balanced against the tendency for moisture accumulation which can favor mold growth and a different source of allergens. This is the principal limitation and makes supply ventilation less practical for residential buildings.(6)

Table 7 summarizes some major advantages and disadvantages of Mechanical Supply Ventilation.

Table 7. Advantages & Disadvantages of Mechanical Supply Ventilation

Advantages

- Outdoor air can be pre-cleaned and conditioned.
- Good control of air flow
- Lower tendency to back draughting

Disadvantages

- Moisture may build up.
- No removal of pollutants at source.
- The challenge of polluted supply air.

BALANCE

In practice, every building is somewhat unique and poses a challenge to optimize indoor air quality by adequate ventilation. The total ventilation system will often combine both extract and supply mechanical ventilation and to some degree, natural ventilation. When combined, the system favors two overall strategies: one that essentially brings in air from the outside which then *mixes* with the air in occupied zones and is then extracted from polluted zones as it were, and alternatively, one that essentially *displaces* the indoor air rather than mixes with it.

With mixing ventilation, heat recovery is possible by using a plate heat recovery unit or some similar 'air to air' system. This allows the incoming air to be pre-heated, which could save some of the additional costs that a mechanical ventilation system incurs, especially in extreme climatic conditions.

The key to displacement ventilation, is to pre-condition the supply air just a few degrees below the room temperature and introduce it slowly near the ground level. It can then spread around at floor level - aided by gravity and possibly intentional diffusers - until it encounters some heat source (like inhabitants, lights or an electrical device) which causes it to rise and be extracted near the ceiling. This potentially energy-efficient ventilation system has become increasingly popular in offices and public buildings, especially in Europe. But it remains challenging.

Modern ventilation systems design has tended to focus on effective control. With strategically placed sensors designed to monitor dominant pollutants of choice, an efficient electronic control module to turn systems on and off, and a modern balanced mechanical system, it is now feasible to control ventilation on demand.

IT IS NEVER ENOUGH!

Ventilation is an appealing strategy for dealing with indoor air pollution. A supply of fresh air from the outside, leading to displacement or at least dilution of the indoor air, and an adequate means of exhaust or removal, would altogether seem to address the problem where and when it does exist. But it is never enough.

You can be sure that any ventilation system will not adequately protect you from the microbes, VOC's, allergens and all the other contaminants which were described in Part One of this book. It will indeed be a useful strategy in any complete response, but it clearly has limitations.

First, there is the energy consideration. When it is very hot, we resort to air conditioning. In winter, we heat our homes and other living/working spaces. In either case, these interventions are as much for our comfort as they are for health preservation. Yet they result in tremendous energy costs. But whether it is to cool down the hot humid air or warm up the cold frigid air, heating and air conditioning costs have become major burdens in today's energy-conscious and energy-strapped environment. That is why in recent times, especially since the energy crisis and the oil embargo back in the 1970's, the focus of architects and engineers, builders and homeowners alike, has been on creating sealed, airtight environments to preserve energy and reduce costs.

Ventilation problems arising from inappropriate design, operation and maintenance can hardly be overstated in our modern 'tight' buildings. It is probably fair to say that in many cases, poor ventilation is the biggest cause of indoor air quality problems. It has even been estimated that 75% of buildings do not get enough outside air, 65% have problems with adequate air distribution to all occupants (since partitions and other obstacles often block supply and return vents); 60% have improper filtration, 75% have inadequate maintenance of the HVAC or other air handling systems, and 45% have microbial contamination.(7)

So what we have 'gained on the swings, we have lost on the roundabout'. Now we have trapped ourselves in polluted prisons that cause consequences to health that we cannot afford. **We have compromised our health to some degree for the economics of energy conservation and the comfort of ambient temperatures and humidity.** Are we crazy or what?

The second challenge we face with increasing ventilation is the fact that often, to bring in more air from outside to dilute or displace the polluted air inside, would make little gain because **the air outside can be polluted in its own way and for whatever reason.** This is true in many areas of contemporary industrialized society. We're no longer isolated in the mountains, or in the open countryside or dwelling on the beach. We live in large urban centers with the wheels of industry turning all about us, as we hustle to and fro' by every means of transport we can imagine. Combustion of fossil fuel, the production of toxic waste and the use of myriads of volatile chemicals and cigarette smoke, all combine to poison the *outside* air and thereby threaten further, everything we try to do by ventilating *inside*.

The third important limitation that should be mentioned regarding ventilation is the failure to adequately address bioaerosols or droplet nuclei which are ever so important in **the spread of infectious disease**. SARS gave us all a scare in recent time but as we mentioned earlier, infections like the common cold and influenza are still prevalent, and airborne transmission is still a major consideration.

In almost any setting, less ventilation will increase the risk of airborne infection and more ventilation will reduce it. That seems obvious. But any standard level of ventilation chosen to prevent infection is inherently a compromise. It will always depend on the conditions of exposure and the level of risk that society is willing to accept for a given cost. In practice, each doubling of effective, well-mixed ventilation reduces the risk of infection by about one-half. Therefore, in some cases, it may be possible to double or quadruple room ventilation depending on baseline conditions. But **such risk reductions of a half or a quarter for disease transmission can hardly be considered satisfactory protection.** And that would be the situation at best. Moreover, it is a general principle that as airborne concentration of organisms are reduced, ventilation becomes increasingly inefficient in reducing their concentrations further. Even the comparatively low level of air contamination still presents a health risk for occupants sharing that air for extended periods of time. This is especially important and true for infections like tuberculosis, where the infection dose could be as miniscule as just *one* inhaled droplet nucleus.

In summary, the strategy of increasing ventilation certainly has an important role to play in combating indoor air pollution and its threatening consequences, but it has clear limitations which guarantee almost invariably, that even combined natural and mechanical ventilation is still not enough.

So that's not a practical solution much of the time. We must go beyond the attempts to **eliminate sources** of indoor air pollution. That was **Strategy #1**. We must also go beyond the effects to adequately **ventilate** the internal environment where we live and work (for more than 90% of our time, remember!!). That is **Strategy #2**. That's where we may choose to begin, but that is still just the beginning.

That takes us to the next strategy among the **FIVE ESSENTIAL STRATEGIES** that will get us to **Fresh Air for Life**! It involves **filtering the air** to remove suspended particles, aerosols and volatile organics where possible. That will be **Strategy #3** to be outlined in the next chapter.

STRATEGY #3

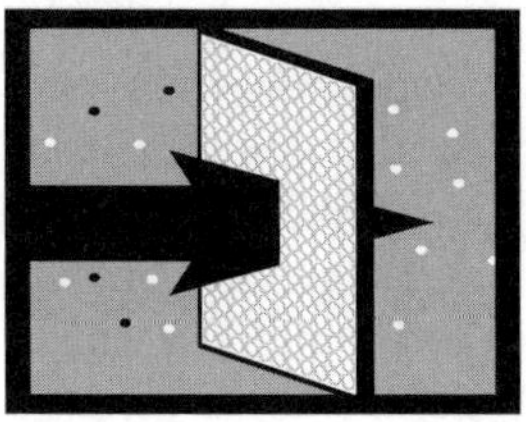

EIGHT

Filter The Air

*"If size were everything, you could filter anything. But **that's not all there is**."*

It should seem obvious that solid or liquid airborne particles, suspended in both ventilated air (from outside) and re-circulated air (from inside) could be filtered out to protect occupants from possible health consequences. It may not be as evident, but gases and vapors that are also pollutants can be adsorbed on to appropriate physical media (by adsorption) or chemical media (by chemisorption), to effectively remove them from the ambient indoor air.

Before we discuss the efficiency and challenges of this filtration strategy, it is necessary to look a little more closely at the nature of suspended particles.[1]

AEROSOLS

There are always particles suspended in indoor air. Some are formed indoors while others are brought in from the outside. The general term *aerosol* is used to describe '**any suspension of particles in air, whether solid or liquid**.' It is therefore a sweeping term that includes mist, smoke, dust, fibers, and of course, bioaerosols such as viruses, bacteria, fungi, algae and pollen. These microbes rarely float around in space by themselves but are usually quick to hitch a ride on some other (larger) traveler going by. Aerosols therefore have quite different chemical and biological characteristics but for the purpose of filtration, it is the sheer physical size that predominates.

Aerosols span a very wide range in size, from about 0.001 microns (micrometers, µm) to about 100 microns. It is difficult to conceive just how small these particles are. They are truly microscopic. The micron unit is so small, it would take 10,000 to make one centimeter. To get the picture, consider that the average human hair is about 50 microns in diameter, or that the dot on the letter (i) here, would equal about 400 microns, and the opening on a common needle could be 800 microns wide. It's no wonder that you cannot see these aerosols suspended in the air about you. But they are there, in the microscopic world where tens or even hundreds of them can sometimes fill the tiny space of a single micron. The smaller aerosols (up to about 10 microns) tend to remain suspended in the air for long periods of time, while the larger particles settle out of relatively calm air in just a matter of minutes. The size ranges and settlement times of common indoor particles are given in **Table 8.**(2)

Table 8. Particle Sizes

Type	Diameter (microns)	Time to Settle 1meter
Human Hair	100-150	5 secs
Skin flakes	20-40	
Observable Dust	>10	
Pollens	15-25	
Mite Allergens	10-20	5 minutes
Common Spores	2-10	
Bacteria	1-5	
Cat dander	1 ± 0.5	10 hours
Tobacco Smoke	0.1-1.0	
Metal/Organic fumes	<0.1-1	
Cell Debris	0.01-1	
Viruses	<0.1	10 days

A typical fraction of indoor air would usually show a bimodal size distribution. What that means is that there will tend to be two groups of particles. ***The fine mode*** accounts for those particles that are smaller than about 2.5 microns. They are *chemically* produced usually by photochemical atmospheric reactions or the coagulation of combustion products from automobile and/or stationary sources indoors. These fine particles can remain suspended for several days or more. They include particles so small that they may only be seen using an electron microscope. The larger particle group constitute **the *coarse* mode** and these particles are mainly generated *mechanically* but tend to remain in the atmosphere for mere hours or even less, rather than days.

Perhaps the most hazardous particles, at least in terms of just size, are those smaller than about 1 micron. They are small enough to penetrate deep into the respiratory tract when inhaled by occupants. They can even sometimes cross over the alveolar lining in the very tiny air sacs of the lung tissue, and end up in the blood stream. This poses a real challenge for practical filtration.

To understand some of the issues involved, we also need to examine what air filters really are and how they actually work.

HOW PARTICLE FILTERS WORK

The most common type of air filters in use today for removing aerosols or particles is the fibrous filter. Less practical filters can be made from foam or membranes.

Fibrous filters are exactly that - filter media that are comprised of long fibers arranged randomly in planes perpendicular to the direction of air flow. Common fibers in use are made of cellulose (wood), glass or synthetics. Most of the filter is actually air space, with the fibers occupying about 1-30% of the filter volume. These fibers are usually much longer than the actual thickness of the filter and tend to fold and lie in or close to the filter plane.

Now for an eye-opener. Perhaps most people would presume than an air filter acts like a screen or a sieve. In other words, when particles strike and try to penetrate, they prove to be too big for the spaces between the fibers and meet mechanical obstruction. That's how most ordinary filters work. You see it everyday when coffee is perked, or tea is strained, or leaves are kept from clogging the eavestrough. Is that your idea of air filtration? You would be mis-guided.

Air filtration is a microscopic process. Particles of micron size are removed by their collision with actual fibers. Once they connect, they are captured and remain attached because of the strong intermolecular forces between the particles and the fibers. **It is not the space blocking the particle, but rather it is the fiber holding on to the particle that retains it there.**

Having grasped that essential difference, you can now appreciate the different filtration mechanisms that are active in an air filter:

- **Diffusion**. Even in the streamlines of air flow, the particles show a tendency to additional random motion (Brownian motion), dependent on air velocity. This random motion can take any given particle off course, (especially a very small one), to connect with a fiber and get detained there.
- **Interception**. This is a direct hit where a fiber lies right in the main

stream of a particle's flight path. There's no dependence on velocity here, and particles larger than 0.5 microns are ideal.

- **Inertial impaction**. Heavier particles (more than about 0.5 microns) can't keep up with the twists and turns of the air stream, especially around individual fibers, so they fall away and then connect with fibers in their renegade path.
- **Electrostatic capture**. Under some circumstances, particles can become charged (intentionally or not) and they will be attracted to oppositely charged fibers (by Coulombic forces) or to uncharged fibers (by image forces). The net effect is that there is electrical attraction, especially at lower velocities.

Particles can sometimes bounce off a fiber depending on its mass and velocity, or the fiber size and the direction of impact. Where this becomes critcal and reduces efficiency, the fibers can be coated with a liquid or adhesive to significantly reduce this bounce and improve retention.

FILTER DEVICES

Mechanical Filters

These filters exploit the first three mechanical processes above and do not require any electrostatic forces. They may be used in central filtration systems as well as in portable units using a fan to force air through the filter. These filters have a characteristic resistance to airflow that increases with time as the filter is in use. Why? Captured particles become part of the filter structure itself. At first this is deep in the filter medium but eventually, these particles begin to bridge across the filter surface causing an increased resistance, a ***pressure drop***, and finally a surface cake. The filter has increased collection efficiency but from an energy standpoint, it becomes an expensive device to operate.

In practice, mechanical filters are produced with different designs. ***Flat-panel filters*** are still common and essentially place all the filter media in the same plane - hence, a flat panel. These usually contain a low-packing density, fibrous medium that can be dry or coated with a viscous substance such as oil to increase particle adhesion. The dry type media might consist of open-cell foams, non-woven textile cloth, paper-like mats of glass or cellulose fibers, wood fill, animal hair or synthetic fibers. A typical furnace filter of low efficiency rating is usually only efficient at removing the particles larger than about one micron.

To increase the particle collection efficiency, one can increase the filter media density by using small denier fibers. This allows less media penetration and increases the filter's screening capacity. Such a filter can efficiently remove particles extending down into the sub-micron range. But this comes at a price. Namely, the increase in media density increases the resistance to airflow velocity through the filter.

To compensate, the most effective approach has been to extend the surface area of the filter medium by pleating the filter. ***Pleated-panel filters*** have an accordion-like appearance and show higher efficiency for two reasons. They increase effective surface area and reduce the relative media velocity.

Bag or pocket filters are also very popular for high-efficiency air filters. They expand as air flows through, to expose all the media. ***Removable filters*** which involve a moving (rolling) medium find only limited application.

Some more modern central HVAC systems utilize automatic, self-renewing cartridge filters which can be cleaned in situ (on line, without removal) by briefly operating the system in a reverse pulse mode. These cartridge filters require less maintenance and downtime. They have a high capacity for heavy concentration of particulates and high efficiency. But they need compressed air to pulse the solenoids; they also tend to be rather bulky and usually require a much larger blower system.

Taking out filters should never be done while the system is running and in any case, the potential threat of hazardous exposure demands appropriate safety precautions such as face masks, gloves and careful bagging of dirty filters.

The simplest form of a filtration device is **the self-contained media collector.** This consists usually of two principal components. The filter section is often a combination of a pre filter and a primary filter, and secondly, a motor and blower system is employed to draw air through the media filters. Dirty air is drawn through the pre filter first to collect the larger particles (such as lint and fibers) that would otherwise clog the main filter. Smaller particles are then collected in a more efficient primary filter section. These latter filters are available in a variety of efficiency ratings and should be selected based on the particular application.

Such media air cleaners have both advantages and disadvantages. They are relatively economical to buy and install, require rather low maintenance overall and provide a large choice of filter efficiencies. On the other hand, filter replacement could prove quite costly as they use a significant amount of energy to drive the higher horse power motor to derive the airflow against the filter resistance (pressure drop).

With ***electronically charged mechanical filters*** (otherwise known as *electrets*), the charge itself increases collection efficiency but has no effect on airflow resistance. But they too become mechanically blocked. The effect is a bit more complex, since the deposited material obstructs the electrostatic effect and therefore reduces collection efficiency.

Electronic Filters

Electronic filters are generally marketed as electronic air cleaners that were formerly referred to as *electrostatic precipitators* (ESP's). The simplest form of electronic air cleaner is the unqualified negative ion generator. They operate by charging the particles suspended in the room air, which then become attracted to and deposited on walls, floors, table tops, curtains etc., where they could cause **soiling problems**.

The modern 'two stage electrostatic precipitator' was invented much earlier by Dr. Gaylord Penny (not Perry, sports fans!) back in the 1930's. Just like mechanical filters, they are still used in both central filtration systems as well as in portable units with fans. In central HVAC systems, they sit adjacent to the main furnace in or near *the return duct*. They collect particles suspended in the air stream by electrical precipitation. Here, the air passes between two parallel plates, between which some equally spaced wires function as high voltage electrodes. When in operation, the high electric fields between the electrodes cause a corona discharge that produces (usually positive) ions which charge the particles and cause them to be attracted to the grounded or (actually negatively charged) collector plates.

These units can maintain a relatively high efficiency rate for removing particles in the submicron range. There is no barrier to flow or pressure drop, but energy is consumed in the ionization process. The airflow remains constant with use, but the particle capture efficiency tends to decline as the charged collector plates become coated with particles. However, cleaning the reusable plates restores the initial efficiency and should be done regularly to maintain adequate performance. **Table 9** lists some advantages and disadvantages of these filters.

Table 9. Advantage & Disadvantage of Electronic Filters

Advantages

- Into submicron range
- Low energy costs
- Constant airflow
- Reusable plates (no filter replacement)

Disadvantages

- Efficiency falls off
- Frequent cleaning
- High initial unit (HVAC installation) cost
- Can cause *soiling* on walls, furniture, etc.

FILTER EFFICIENCY

Clearly, filters have a job to do. They should remove as many of the air contaminants as possible and hopefully, at minimal cost and on a reliable basis over time. Their efficiency, or how well they get the job done, can be actually defined and measured in a number of ways. Intuitively, the efficiency should reflect the degree of particle or other contaminant removal. This can be measured, at least in a comparative sense, by standard laboratory tests that compare how a standardized synthetic contaminant dust is removed by any filter in a controlled, repeatable way. In reality, the applied practical conditions may be different from the controlled environment used for such testing, but at least comparative rating of different filters is achievable by such methods.

But there are other ways of measuring efficiency of air filters. Mechanical filters in particular, have a marked *pressure drop* that provides resistance to airflow which in turn tends to increase *energy consumption* and therefore cost. As a result, it is possible to rate filter performance in terms of pressure drop or energy demand. Yet a third consideration is the *lifetime* of the filter under applied conditions. As the filter is used, how will its performance vary over time? What maintenance will be required? Is cleaning feasible and can that be automated? Clearly, answers to these questions provide a different rating consideration.

The gold standard of air filters today is called a HEPA filter which stands for *High-Efficiency Particulate Air filter*. These filters have a rather interesting history. They were first developed during World War II by the U.S. Atomic Energy Commission to fulfill a top-secret military mission. As part of the Manhattan Project related to the development of the atomic bomb, they needed to find an efficient way to filter radioactive particulate contaminants. Needless to say, the primitive HEPA filter technology was declassified after the war and then allowed to enter commercial and residential use. Those early HEPA air filters were nothing like the compact, more efficient filters commonly available today. Modern HEPA filters are made of submicronic glass fibers in a thickness and texture very similar to blotting paper. HEPA filters use a powerful blower to force the air through a very tight membrane to achieve the high efficiency particulate filtration. They have found limited application in central HVAC systems but are quite common in vacuum cleaners, portable air cleaners and other stand-alone units for residential and commercial use.

The biggest advantage of the HEPA filters is that they are very efficient in the filtering of air that passes through the filter and they can filter down to 0.3 microns. But they do require frequent filter changes and can act as a breeding ground for bacteria, mold and fungus. They do not remove odors, gases, pesticides, viruses and many bacteria. They are also utilized for more specialized duties in the nuclear, electronic, aerospace, pharmaceutical and medical fields. HEPA filters generally have higher energy costs and significant replacement costs as well.

In houses and offices all across the industrialized world, HEPA filters and others are now in common use for air filtration where they routinely remove particulate pollutants such as combustion smoke (including environmental tobacco smoke components), pet dander, pollen and dust that can aggravate allergies and cause other respiratory problems. Such filtration is at least a first line defense against indoor air pollution in terms of removing several dangerous contaminants from the air. However, as we shall see later, this is by no means the last word since the process of filtration itself has its own intrinsic limitations and other strategies of purification are necessary for a more complete result.

Using the standardized laboratory methods referred to earlier, it is possible ***to define a HEPA filter*** **as having a minimum efficiency of 99.97% on 0.3 micron particles.** If the aerosol concentration is measured by counting particles according to size, both upstream and downstream from the filter, one can truly measure by some standard tests, the removal efficiency for each particle size. Both optical particle counters and aerodynamic particle counters are typically used to quantify the test dusts which are usually made of synthetic oil aerosols.

Perhaps a more practical way to measure efficiency of say an air cleaner, is to monitor the effect of cleaning room air in a standardized test. Here the efficiency is a product of *both* the **filter efficiency** itself ***and*** the **flow rate** of the air through the device. This allows for a new efficiency term: **the clean air delivery rate (CADR).** This is the product of both terms.

$$\textbf{CADR} = \textbf{Filter Efficiency} \times \textbf{Flow Rate}$$

Let's illustrate. If an air cleaner has a filter efficiency of say 90% and a typical flow rate of say 100 cfm (cubic feet per minute), then the CADR would be, 90/100 x 100 = 90 cfm. **CADR would measure the volume of air being 'cleaned' per minute when the device is in use**. In other words, the CADR may be defined for a given contaminant or for overall 'cleaning efficiency' by measuring the decay rate of aerosol concentrations when the device is applied in a defined chamber. Typically, measurements are made in standardized chambers using fine dust, tobacco smoke or pollen as test aerosols.

PRACTICAL CONSIDERATIONS

But the ability of an air cleaner to clean a room *effectively* will be influenced by other factors, apart from the CADR just mentioned. That is clearly the main factor to consider, but one should also pay attention to the size of the room; the air exchange rate going in and out of that same room, and the quality and composition of the entering air. Other factors would include: the inside sources of contamination; the nature of those contaminants; other mechanisms/strategies for removing them; the airflow patterns of ventilation; the opening and closing of windows and doors, and obviously the quality of the air cleaner itself. In the end, the results of any attempt to clean the air with a portable air cleaner will necessarily reflect the effect of all these factors taken together.

You might be wondering: **why would anyone use a portable air cleaner in the first place?** There are very valid reasons for doing so. Certainly in the absence of an HVAC system, local air treatment may be essential. It may not be always possible to install higher efficiency equipment in older environmental systems. Stand-alone units may also find appropriate application for rapid removal of environmental tobacco smoke (ETS), for example, in a hotel or cruise ship seeking to satisfy the needs of discriminating guests, or in any isolated or confined space where rapid results are desired.

Portable air-cleaning devices often contain, in addition to a filter or electrostatic precipitator, a fan to increase air flow, some activated charcoal for absorption, and a UV light source for further purification, as we will see later.

Various types of filters are used in a number of ways for different applications in both homes and offices. The key consideration of course is always to effectively remove contaminants from indoor air with safety, reliability and minimal cost. When considering the choice and deployment of an air filter, it is useful to address a number of critical questions:

- *What is the purpose or objective at the point of application?*
- *What type of particles are in the air?*
- *What characteristics of the filter are needed?*
- *What design characteristics of the system should be included?*

Answers to these types of questions will allow for the best possible application of the ideal filter for each particular situation.

ACTIVATED CHARCOAL FILTERS

We have seen how particles and aerosols are removed from polluted air by the use of mechanical and electrostatic filters. However, many airborne contaminants are present in the form of **gases and vapors** which would otherwise pass through such filters and therefore require alternative or supplementary procedures to remove them and avoid the consequences of their inhalation.

Of the variety of different procedures that have been researched and developed for removing unwanted vapors and gases, the process of physical and chemical *adsorption* has found wide application.(3) Perhaps that is due to the simplicity, efficiency and comparatively low economic cost. (Just in case you missed the subtle distinction, the general process of **adsorption** takes place on solid surfaces, in contrast to **absorption** which involves dissolution in liquid media).

For physical adsorption, activated charcoal has become by far the most popular medium in a wide variety of indoor environments. This unique material is formed in a two-step manufacturing process. First, a carbonaceous material like coconut shells, wood, peat or coal, is charred in the absence of air to remove the hydrogen and oxygen as water vapor and leave a block of carbon. This is then 'activated' in a second step by slowly oxidizing parts of it in the presence of steam to create a network of pores that extend throughout the bulk of the material. These pores vary in size from less than 2nm (*micropores*), up to 50 nm (*mesopores*) and above (*macropores*). It is the smaller pores (roughly under 5nm) that are active in adsorbing the airborne vapor contaminants. In effect, the **activated carbon provides a very large surface area for adsorption (more than 1000 square meters or almost half the size of a football field for each gram of material)**. Remember, it takes almost 30 grams to make just one ounce. About half of all the carbon atoms form the total internal surface of the pores that permeate this activated material.

Activated carbon is available in many commercial grades for different applications, from common barbecueing to highly specialized purification. The American Society for Testing and Materials (ASTM) has therefore developed some basic standards for rating the quality of activated carbon based on density, particle size distribution, total ash and moisture content, carbon tetrachloride activity that measures micropore volume, and what is known as 'ball pan hardness.' This last characteristic is a measure of attrition, or the tendency of the carbon to settle out as a loose powder from the induced vibrations as the air passes through it. (It dervies its name from the way it is measured ... you can use your imagination!)

The activated charcoal filter is typically employed as an adsorption bed through which the air is passed and the mixture of vaporized contaminants compete for the many active sites. The efficiency of adsorption will depend on the contact time in the filter, which is longer for a thicker bed and a lower air flow velocity. Another important criterion is the opposing pressure drop across the filter, which demands an energy cost. This on the other hand, is higher for the thicker bed and at higher velocity. Therefore, one must always balance the efficiency against both the cost of the carbon filter and the energy cost in operation.

It is interesting to note that, unlike mechanical filters for particles, **the pressure drop across carbon filters does not change as the contaminant is adsorbed.** Therefore, it does not deteriorate as quickly or form a surface cake. Its predicted lifetime must therefore be assessed by the manufacturer and the filter should be changed routinely. A good guiding rule of thumb is to assume that it takes about 4.5lb of activated carbon in a filter to be effective at 'cleaning' about 1000 cubic feet of air per year.

Activated charcoal itself is not effective in removing several important indoor air contaminants like sulfur dioxide, low molecular weight aldehydes and organic acids (including formaldehyde and formic acid), nitric oxide and hydrogen sulfide. But it can be impregnated with different chemicals which do react with these contaminants. This provides for chemical adsorption or chemisorption, a process which depends much on contact time, relative humidity and temperature.

Another popular matrix for chemisorption from air is made by impregnating alumina with the more reactive potassium permanganate. This medium is superior for removing the most reactive contaminants like nitric oxide, sulfur dioxide, formaldehyde and hydrogen sulfide. Sometimes it is best to combine the activated charcoal and the impregnanted alumina in a 50:50 mixture.

BIOLOGICAL FILTERS

As we have emphasized repeatedly, a major component of air pollution consists of microorganisms in the form of bacteria, viruses, fungi, mold and algae. These microbes are known to transmit disease as they follow air currents from one infected person to another vulnerable co-inhabitant. To reduce this threat, the goal of air disinfection would be to reduce the concentration of infectious particles by *dilution* (through ventilation), by *removal* (through filtration) or by *inactivation* (through a variety of ways we shall outline in later chapters).

Filtration is indeed a reasonable mechanism for removing infected droplet nuclei from indoor air.[4] This technology was first studied by a Harvard sanitary engineer named William Wells, back in the 1930's. He was ably assisted by a young medical student named Richard Riley and together they were able to demonstrate that **the real infection carriers were formed as dried residua**

(ie. droplet nuclei) from almost instantaneous evaporation of larger respiratory droplets.

The net effectiveness of filtration as a procedure for disinfection must reflect the presumption that airborne transmission is a (if not, the) major route of dissemination. One must also presume that the area to be filtered is the actual site location for the spread of the infection of concern. In other words, it would be pointless to target air disinfection by filtration in an area, if the infection were not truly airborne in transmission, or if there were other important sites of transmission excluded from the same area.

The droplet nuclei that are the most effective agents of infection transmission are not the same as the large particles or aerosols that are released into the air by coughing, sneezing or other respiratory maneuvers. Those latter emissions tend to be large, averaging about 10 microns in diameter when exhaled. They are most likely to settle rapidly on to surfaces where they will dry and become part of the household dust. Even if they remain viable and become re-suspended in the air with dust particles, those are again relatively large, so they tend to resettle quickly too. However, it is the smaller droplets that contain viable microorganisms, which evaporate fairly quickly down to about 1 to 3 microns. They form droplet nuclei that remain airborne indefinitely as they are swept along by ordinary room air currents. Those become the major agents of airborne infectious disease transmission when they are inhaled.

The principle of using filtration to disinfect indoor air is therefore to attempt to trap these smaller droplet nuclei. This obviously requires high quality filters. But the elite high efficiency particulate air (HEPA) filters are designed to trap the more penetrating 0.3 micron size particles. They will do the job, but even somewhat less efficient filters should still be highly effective for reducing airborne infectious particles.

The challenge of the filtration approach is not so much in the filter itself as in the intrinsic requirement to pass the air currents containing the droplet nuclei through the filter itself. Just as we observed earlier with ventilation, even if one could double the rate of well-mixed air passing through the filter that would only reduce the risk of infection by a factor of just two. The issue remains: how to re-circulate the air adequately so that the important droplet nuclei can get trapped in the filter before they are inhaled by someone else in the same occupied space.

Then there's always the tail of the particle size distribution, below the 0.3 micron level. These *fine mode* particles (nuclei mode) can also carry microorganisms which are much smaller. Typical dimensions for bacteria would be in the 0.05 to 0.7 micron range, and for viruses, even smaller at less than 0.01 up to about 0.05 microns. They can attach to almost anything, even the fine nuclei. These much smaller nuclei will still evade the elite HEPA filters under the best of circumstances and other disinfection methods are therefore necessary for completeness.

HEPA filters have indeed been used in central ducts of HVAC systems to trap recirculating droplet nuclei. More commonly though, stand alone room units with HEPA filters and other purifying modalities which we will discuss later, have been used effectively in more localized high risk situations. The key is to position any such device as close to the source as possible (if known), and away from any air intake.

Air filtration does have such intrinsic limitations.

THE PLAIN TRUTH

Air filters are available commercially in all shapes and sizes, and in a variety of devices, but more importantly, there's great diversity in their performances, often at variance with advertised specifications, claims and expectations. It is a challenge for the consumer to find the plain truth.

Several key considerations need to be clarified and emphasized.[5]

1. ***The term 'efficiency' is much abused.***

You would think that the efficiency rating of a filter device or apparatus is the best indicator of its quality and performance. Not so. The rated efficiency pertains to a particular particle size and usually that pertains to the filter itself. For example, a HEPA filter with a 99.97% efficiency rating refers to the filter performance with particles of 0.3 microns size or larger. That is a value-on-paper. The language sounds like 'any device with a HEPA filter is removing 99.97% of any and all air contaminants'.

However, things that sound too good to be true, often are. They cannot be just taken at face value and acted on. They need to be analyzed and justified, at least to the discriminating consumer. In a given device where such a filter may be used, the true efficiency could be as low as 80% due to leakage, damage, contamination, humidity and other factors. Note also that the effectiveness of the device in use will depend on the airflow rate, ventilation, short-circuiting of airflow, possible intake contamination, aging of the filter, and more.

What's the bottom line? **You should never be deceived by the exaggerated statements about an *air filter* device** that "*effectively removes 99% of all airborne allergens*", or how about this one, "*effectively scrubs the room free of all air pollutants*". Such descriptions would suggest that these devices can virtually remove all the unwanted contaminants from the air in a normal indoor environment. However, it's just not that simple.

In some high-risk situations, like in clean rooms and hospitals, professional HEPA filter systems are designed and maintained for exceptional performance. The fact that a HEPA filter is used - or even worse, a *HEPA-type* or a *HEPA-like* filter is used - is no guarantee of similar performance in more common applications. These 'imitation' filters may have efficiencies as low as 55% or less at 0.3 microns, even though they have the same physical style. Plus, remember that all filters are not HEPA filters. They're not all created equal.

2. Efficiency is not everything.

This is so obvious and yet it is not. A device that is designed as an air filter will have many components and different challenges, especially the stand-alone room units. Air has to circulate, for only the air that is able to pass through the filter will indeed have even a chance at being filtered. It is common practice to incorporate a fan driven by an electric motor. As the fan rotates, it draws air through the unit (and through the filter, in particular). The free-flow air handling capacity of the motor and fan should never be confused with the actual air flow rate when the filters are in place, i.e. when the fan is in use in practical application. Filters provide a resistance to flow and the difference in rates can be very significant. Therefore, estimates of the number of air changes per hour for a given room can be grossly exaggerated.

In addition, when air is re-circulated, it is important to consider the actual patterns of airflow. Unless the unit can be effectively placed near the source of contamination, it is more than likely that thorough mixing of filtered and unfiltered air may not be taking place. There could be *dead space.* So a simple volume calculation may not be an accurate estimate of the real times required to effectively remove almost all the filterable contaminants.

In many central HVAC filtration systems, *filter by pass* is also a common problem. It occurs when air goes around the filter because of a poor fit, improper sealing in the framing systems, missing filter panels, or leaks and openings in the air-handling unit between the filter and the blower. Simply improving filter efficiency without addressing these potential by-pass problems could be of little benefit.

Outside building walls are also quite leaky and the effect of negative indoor pressures (usually warmer air) allows significant *infiltration* of outside air. If that air is polluted, then the consequences are obvious. Steps to minimize such infiltration can then have significant impact on the net efficiency of efforts to improve indoor air quality.

3. Gas phase filtration has its own requirements

The efficiency of a high quality filter is usually defined by its ability to remove particle contaminants. This has no relationship to the removal of gaseous products and associated undesirable odors. This latter challenge is left to the adsorption or chemisorption filters we described, of which activated carbon is by far the most common. But that cannot be taken for granted. When gaseous contaminants are important or even predominant, it may be the efficiency of the activated carbon that limits the performance of the air filter most.

It is well known that activated carbon in its granular form is eminently effective for removal of many gaseous contaminants. But yet most air purifiers on the market today utilize an inferior product consisting of carbon fiber pads which are only impregnated with activated carbon dust. The performance is not comparable. In addition, special adsorbents are required for low molecular weight gases like hydrogen sulfide, ammonia or formaldehyde. The cheaper natural mineral known as zeolite is an ineffective substitute for good quality activated carbon.

4. ULPA filters can be seductive

ULPA is an abbreviated acronym for Ultra Low Penetration Air. The ULPA filter is supposed to have an even higher efficiency than the HEPA filter. Typical ULPA filter media have an efficiency rating in excess of 99.999% at 0.12 microns. They can, in principle, remove more particles than the HEPA filter media and as such, find application in modern clean rooms like in hospitals, electronics manufacturing, pharmaceutical labs, and so on.

But in normal applications where units containing ULPA filters are used to purify indoor air in homes and offices, this ultra fine filtration could be a distinct disadvantage. The ULPA filters tend to be much denser than HEPA filters and that reduces air flow significantly. That reduction could be as much as two to five times less. Therefore, there are fewer air turnovers per hour in any typical room. All other things being equal, most room purifiers with HEPA filter(s) would probably be more effective in reducing the particle concentration in any typical space, than the equivalent device using ULPA technology.

5. Air filters should stand up over time

Naturally, the real test of any air cleaning device is its effectiveness after extended use. This is true of any filtration system because, by definition, filters tend to become blocked and inefficient, unless their protection is carefully designed and maintained. Air filters that show excellent results, but only do so for a few hours or even days of usage, will have built-in obsolescence and extravagant economic costs. Only long-term performance should be acceptable.

The HEPA filters represent the optimum filter technology for removing particle contaminants from air today. They trap particles down to the fractional micron range. But they need to be protected upstream from the much larger air particles that would otherwise clog them quite readily. The use of effective pre-filters in HEPA based air purifiers is almost mandatory to protect those filters, as well as the activated carbon filters which have pores that can indeed be clogged by dust particles. Frequent clogging of filters is both inefficient and expensive.

Good maintenance is also required. In central HVAC systems, filters should have permanently installed indicators of pressure drop that alert the technician or homeowner that a filter change is needed. In the case of more modern, automatic, self-renewing filters, there should similarly be indicators when the media supply is exhausted. Many homeowners make it a common practice to change their main HVAC filters at the beginning of each winter season. Problems with odor or biological growth may require more frequent filter changes.

It cannot be overemphasized that changing your HVAC air filters is **NOT** the only maintenance required to ensure good quality air indoors. A complete maintenance program of just the central HVAC system requires making sure that, among other things, condensate drains are clear, ducts are attended to appropriately, and also, humidifiers and electrostatic cleaners are adjusted and functioning properly.

All air filters - mechanical, electrostatic or activated carbon - found in stand-alone air cleaning devices should also have manufacturer's maintenance schedules based on actual hours of usage. All filters degenerate with time and wise application must necessitate consistent adherence to a well-defined maintenance program. The cost of good, regular maintenance will be more than justified by the better performance of the air purifying device; the cleaner and more pleasant quality of indoor air, and the healthier and more enjoyable lifestyle that is afforded to the residents or inhabitants where these devices are deployed.

Nothing less will do.

We have now outlined the first three of the **FIVE ESSENTIAL STRATEGIES** that we have proposed to find an effective Solution to the Problem of indoor air pollution. Clearly these three steps will go a long way to reducing the levels of indoor air pollutants. But they all, even together, have serious limitations which we have identified along the way.

When you get right down to it, air purification must involve dealing with impurities, both physicochemical and biological, that *cannot be completely eliminated* at source; *cannot be effectively removed by ventilation* in a convenient or cost-effective design, and *cannot be efficiently filtered* out of the air for a variety of reasons. Sooner or later, the conflict with the enemy will demand face-to-face, hand-to-hand combat. In other words, we have to target the enemy, disarm and destroy his destructive capability and finally put him out of commission.

This is an indoor pollution war! We go next, right to the battle front!

STRATEGY #4

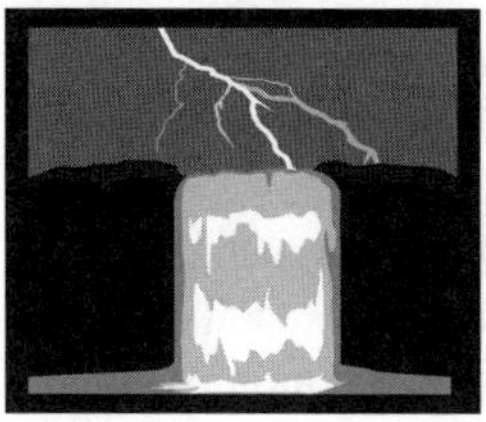

NINE

Purify The Air

"It is always a challenge to improve on ***what nature does best.****"*

Purifying the Air is the real battle front where we attack contaminants in the air on their own turf and neutralize them. That's where the final battle against indoor air pollution is either won or lost.

Nature fights this same battle outdoors each and every day and after all this time, nature is still winning. 'He, she or it' (however you personify nature) must be doing something right. The inescapable track-record proves that. Therefore, it would appear self-evident that to copy or imitate the best design of nature is always the wisest approach to solving any common problem. Air pollution is no exception. After all, despite the complexity of nature with all its diversity and perennial activity, and despite the irresponsible and sometimes inescapable pollution of the environment, somehow by Intelligent Design, the air we breathe (which is our natural habitat, you recall) remains conducive to life, and by and large, to even a reasonably healthy existence.

How is it that the atmosphere withstands the forces of nature - the **normal** cycles of sunrise and sunset, sunshine and rain, summer and winter - and the more *unusua*l environmental shocks like tornadoes, thunderstorms, lightning and whatever atmospheric changes we neither perceive nor appreciate? How is it that despite the constant man-made pollution of the atmosphere with industrial emissions, automobile exhausts, volatile chemicals and much more, the atmosphere as a whole has not become toxic enough to make the planet a barren wasteland? At least, not yet.

Clearly, by Intelligent Design, there are natural counter measures taking place in the atmosphere that help to neutralize or destroy many of these pollutants and so maintain a livable environment. In fact, some of these natural phenomena that seem so threatening and destructive - like lightning and thunder - are part of the overall scheme of nature to protect the atmospheric equilibrium. They are blessings in disguise, if the truth be known.

We now understand that there are natural anti-pollution forces at work outdoors. Three of these have been studied and understood to some degree, namely: Ozonation, Ionization and UV Radiation.

First, we know for example, that normal sunlight (and also, thunderstorms and lightning) converts normal oxygen into the more reactive ozone species. As we have already noted, ozone goes a long way to removing hydrocarbons and other organic molecules from the atmosphere by oxidation. That is the process we refer to as ***Ozonation***.

Secondly, we have also learned that, unlike the high vacuum of space where atoms and molecules are present in excited states and possess electric charges, most matter on earth and in its atmosphere are un-ionized. However, natural high-energy radiation and other forces produce some low levels of ionized species in the atmosphere. These ions are reactive and help to physically remove particles from the air as well as to chemically neutralize volatile organ ics. That is the process we refer to as ***Ionization***.

Thirdly, we also understand that solar radiation has a high-energy tail in the ultraviolet region of the electromagnetic spectrum which proves lethal to microorganisms. Common microbes do not survive very long in the outdoor air because the high-energy ultraviolet from the sun is just right for the destruction of DNA, RNA and important life proteins. The die-off rate in the outdoors varies from one pathogen to another, but can be anywhere from a few seconds to a few minutes for an over 90% kill of viruses or contagious bacteria. They fall victim in the process to the lethal force of ***UV Radiation***. Humans are not at risk because this UV-C radiation, as it is called, does not penetrate the skin very well.

That is nature's way. Where nature cannot *eliminate* or avoid pollution sources, where nature cannot *ventilate* the air because the atmosphere is surrounded by a more rarified ether, and where nature cannot *filter* out fine particles at will - **nature resorts to purifying the air by those three methods just mentioned: ozonation, ionization and UV radiation.**

When the problem of air pollution is moved *indoors*, and when the best of nature's anti-pollution methods are kept *outdoors* - then we must learn to copy that Intelligent Design effectively. And that, scientists have done. Therefore, after we have taken control measures to ***eliminate*** known sources of indoor air pollution, to ***ventilate*** our sealed homes and offices within the limitations of cost-efficiency and comfort, and finally to ***filter*** as many of the contaminants as it is practical and feasible to do - then we must resort to use the same three methods we learned from nature outdoors to ***purify*** the air indoors. We should resort at least to Ozonation, Ionization and UV Radiation also.

It is always very hard to beat nature at what it does best. After all, nature is nothing less than the product of Super Intelligent Design. So this becomes our **Fourth Major Strategy** for dealing with the problem of indoor air pollution. We will seek to purify the air by copying nature's way.

We will discuss each of these conventional air purification technologies in turn, throughout the rest of this chapter. It's still not the final answer, but we will get closer to the synergy of space-age technology as we do. However, you will have to wait for more. We will eventually get to **Fresh Air for Life.** Be patient.

Let's begin with Ozonation.

OZONATION

From earlier chapters, it is clear that ozone has earned its rightful place as an effective purification treatment for both *water* and *food.* Ozone generators have been used for over a hundred years for purification and sanitation purposes. It's properties as a bactericide, viricide, fungicide and deodorizer are well accepted by the scientific community. It should be no surprise that ozone has also been effectively applied in air purification as well, but at much lower concentration levels. And that is the key! There is a *low-level* ozone to be exploited in purifying air by **Ozonation** and there is a *high-level* ozone that could be hazardous to health and therefore should be carefully controlled by proper operation of any air purification device, in conformance with the manufacturer's instructions. There is a big difference!

Ozone Levels

You recall that ozone is widely used in the heating, ventilation and air conditioning (HVAC) industry for duct cleaning and disinfection. It is also used by disaster restoration companies. These professionals utilize ozone to disinfect sick homes, destroy mold, mildew, fungi or smoke from fire damage. However, they typically utilize dosage levels in the 1 to 5 ppm range for this purpose, well in excess (by 20-100 times) of recognized standard safety levels. At these high concentrations, it is definitely mandatory that homes and offices be evacuated. It is perhaps a good thing that just the smell of ozone would make any rational person *duck for cover* … fast! They will be out of there! That's not an issue.

Several years ago, a few companies introduced ozone technology as a type of air purifier device to be used in occupied spaces. You can routinely find ozone-based air purifiers in major commercial food processing plants, for example, where they are routinely used to reduce airborne microbials to help protect our food supply. These systems are routinely approved by the U.S. Department of Agriculture and the Food and Safety Inspection Service (FSIS). The FSIS is particularly concerned about worker safety and often requires *ozone air monitoring* to assure levels do not exceed 0.04 ppm.

As we pointed out in Chapter 5, there is no universal consensus as to when ozone levels become harmful to humans. Based on available research, regulatory agencies such as the U.S. FDA and EPA have set safe level standards at 0.05 ppm for continual exposure. The Occupational Safety and Health Administration (OSHA) has set 0.1 ppm as a maximum for the average level during 8-hour (work day) exposures. Other agencies and organizations have similar recommendations,

Low-level ozone systems have been installed in almost every major type of food group production facility, as well as most Fortune 500 food processing facilities. These companies have routinely performed exhaustive studies on the efficacy and safety of this equipment before purchasing and implementation in their respective operations.

One of the most heavily debated issues is the effectiveness of ozone for disinfecting air, because many antagonists believe that it is only effective at very high levels which are unsafe for human exposure. But on the other side, many outstanding scientists have held that ozone is effective at destroying bacteria and other microbes at sufficiently low levels which are believed to be safe for human exposure.

The truth is that **ozone is an effective antimicrobial agent at low levels, but its effectiveness is dependent on a number of factors.** Just like many other oxidative chemicals, disinfection rates for ozone are dependent on the type of organism, treatment time, temperature, relative humidity, pH, the presence of ozone-oxidizable materials, the tendency of microorganisms to form clumps, and the proficiency of the ozone contractor. Humidity is one of the most crucial factors with studies showing that as an airborne antimicrobial, the effectiveness of ozone is optimized at levels higher than 45% RH (relative humidity). Testing at levels below this range gave inconclusive results.

There is only a limited amount of published data available on the research done with ozone at low concentrations. In **Table 10**, Kowalski and others compiled data on ozone use for reducing bacteria and virus populations reported by previous investigators.[1]

Table 10. Ozonation of bacteria and viruses in air

Test organism	Ozone (ppm)	Time (sec)	% Reduction	Investigators
S. salivarius	0.6	600	98	Elford et al. (1942)
S. epidermis	0.6	240	99.4	Heindel et al. (1993)
pX174 (virus)	0.4	480	99.9	De Mik (1977)

Studies conducted by Midwest Research Institute, using 0.05ppm also showed reductions in five different pathogens.(2) Reductions in *E. coli, Staph. aureus, Salmonella choleraesuis* and *Penicillium chrysogenum* populations were between 30% and 70% following 6 to 24 hour exposure. Reductions of *Candida albicans* were even greater at 90%.

Low-level ozone is already finding very significant application as an antimicrobial for air treatment in different parts of the world. For example, the Chinese Government used low level ozone (0.02 ppm) systems, among other things, to help reduce the risk of SARS in some subways and buses. Low level ozone systems are also embraced by many hospitals and nursing homes for control of odors, airborne viruses, bacteria and mold.

One of the world's largest cruise ship/theme park and hotel chains conducted a one year independent study and determined that low levels of ozone and other oxidizers did in fact reduce airborne mold, bacteria and odors. A very important test result from one of those independent controlled studies showed that it is in fact possible to achieve a six log reduction (99.9999%) of the Norwalk virus. This is the virus that has given a bad name to the cruise ship industry and that often requires them to turn back ships due to large-scale outbreaks of Norwalk virus infection. The low-level ozone technology, enhanced by other strategies to be described later, is now being employed in hotel rooms, cruise ships, animal care facilities etc. to reduce levels of air contaminants.

A recent application of this technology is being pursued at the national level as part of the *War on Terrorism*. Ozone, at levels of 0.02 ppm and combined with other oxidizers, is proving very useful for Homeland Security Projects to help fight the prospect of bioterrorism. The US National Laboratory is developing products to destroy viruses and bacteria that could also be used to combat biological weapons of mass destruction.(3) What is it that these researchers know that the "skeptics of ozone use" do not themselves know?

Ozone may indeed play a vital role in the prevention and control efforts for bioterrorism and biosecurity threats. According to Dr. Sherwood from the University of Georgia, what is most needed now is the strategic application of chemicals to combat disease that will be used at low application rates, pose minimal environmental risk, and have a low potential for the development of pathogenic resistance.[4] Ozone technology has the potential to meet all these criteria. Could that be part of the answer to the last question?

That should begin to satisfy the skeptics of this technology but more work will have to be done to quantify its effectiveness in different situations. On that point, it is noteworthy that **Ozonation is usually not used in isolation today.** In the next chapter, we will update the application of this technology in space-age devices that create synergy with other strategies, and exploit the value of ozone as an oxidizing agent, while minimizing the actual exposure to ozone by those who occupy the spaces in question. We have certainly come a long way.

Ozone Sources

Ozone produced for commercial application is usually generated by one of three different methods: corona discharge, UV radiation and electrolysis.

The ***corona discharge***(CD) method is probably the most widely used in air purifiers. Corona discharge occurs when a surface or point source contains an excess of electrons at sufficiently high negative potential that allows surrounding air molecules to take up electrons and become ionized. As the ionized atoms undergo changes in energy level, the gas emits light so the discharge is frequently visible as a faint blue glow.

This CD method uses oxygen (or dried air) passed between two closely spaced electrodes under a nominal applied voltage of about 10kV. The electron velocity passing between or across these electrodes is extremely high and it generates enough excitation to electrically separate the paired atoms of the oxygen molecule, which then go on to attach to other oxygen molecules to form ozone.

Conceptually, this is not entirely unlike the natural formation mechanism by lightning and thunderstorms which we mentioned previously. High energy breaks the oxygen molecules apart to form highly reactive oxygen *atoms* that readily attach to other neutral oxygen *molecules* and the product is this 'superoxygen' or 'active oxygen' that we know as ozone.

Corona discharge ozone generators that are currently available commercially are capable of producing ozone in the gas phase at levels of 1 to 5% by weight in air and up to 14% by weight in high purity oxygen. These are high levels of ozone, difficult to minimize and to regulate. In addition, this discharge method is quite costly to maintain and in the use of air itself, the nitrogen present gives rise to contaminants like nitric acid and nitrous oxide which have associated corrosion problems. It is therefore not the normal method of choice for air purification by ozonation. But corona generation is still widely used in industrial ozonation water systems, for example.

In the ***UV radiation*** production of ozone, the process is similar to the photochemical production which occurs in the stratosphere. Here again, oxygen atoms formed by the photo-dissociation of oxygen by short wavelength UV radiation from an excited krypton gas lamp, react with oxygen molecules to form ozone. An advantage of using UV radiation to produce ozone is that ambient air can be used efficiently as a feed gas. The low concentrations achieved by UV radiation may not work well for water applications but they are ideal for air treatments where high concentrations are **not** required. It may also be the method of choice because it does not have the high maintenance cost and high voltage of the corona method, nor does it produce nitrous oxide and nitric acid.

Perhaps for completeness, we should mention that ozone can also be formed unintentionally from the effect of UV radiation in the immediate surroundings of electrical equipment utilizing lamps with UV emission such as photocopiers, projection equipment and laboratory spectrophotometers. It may also be produced in the areas of welding arcs and some electric motors, such as you find even in your refrigerator, washer and dryer, HVAC blowers and fans.

High current density ***electrolysis*** of aqueous phosphate solutions at room temperature also produces ozone gas. The relatively low ozone concentration produced by this method normally makes it less effective at destroying microorganisms. Alternatively, electrolysis of sulfuric acid can produce very high ozone concentration in oxygen when a well-cooled cell is used. Ozone produced by UV radiation, in comparison, is produced at the ideal concentration required to destroy airborne bioaerosols and volatile organic compounds, yet it can be controlled to avoid the potential health risks with human exposure. Under these circumstances, ozone will deliver.

What will Ozone do?

Ozone is produced intentionally in many quality air purifiers using the corona discharge method. They simulate the effect of natural lightning with a corona electric arc or spark. Strong ozone is produced by converting the oxygen (O_2) to ozone (O_3). This can be used for rapid or more thorough cleaning of the air when human exposure is not an issue. For example, this may be a supplementary setting for use in an '*Away Mode*' when a space is not being occupied. In other devices, ozone is produced incidentally by the effect of ultraviolet radiation and tends to be at relatively low levels.

The ozone produced in either case is an oxidizing gas that travels throughout the available room (space) and oxidizes almost all organics. Ozone can neutralize most odors and certain gases. As we already noted, it is also capable of destroying microorganisms. The ozone units can be installed in central HVAC units or in any room. They do not reduce airflow.

Two particular effects of ozone when used as an air purifier should be described in a bit more detail: first, as a deodorizer, and then as an antimicrobial agent.

Odor problems originate from numerous sources: bacteria, tobacco smoking, fumes from chemicals, cooking, fireplaces and pets. Odors can be big problems when they are affixed to clothing, furniture, fabrics or carpets. Mold and fungus contamination are another major source of unpleasant odors. Damp spots around humidifiers, attics and crawl spaces under homes, basements, bathrooms, house plants, air ducts, damp ceilings and walls, wet carpets and windows - all these are sources for contamination. Mold creates a musty , stale odor which can be both an annoyance and a health issue to those suffering from allergies or asthma. Condensation from steam and poor ventilation is the biggest cause of mold in bathrooms and around clothes dryers or stoves when they are not properly vented to the outside.

Ozone is one of the best technologies available today for odor removal. Airborne ozone has been used effectively in removing odors from previously occupied homes, including odors from pets and molds. It is still debated by some researchers as to how ozone works on odors: by masking them, eliminating them, or both. However, the effectiveness of ozone at eliminating unwanted odors is well documented.(5) The main theory behind the ability for ozone to remove odors is quite simple. When ozone comes in contact with organic compounds or bacteria, the extra atom of oxygen destroys the contaminant by oxidation. Ozone decomposes to oxygen after being used, so no harmful by-products result.

Ozone is one of the best technologies available today for odor removal.

Ozone will neutralize virtually all organic odors, specifically those that contain carbon as their base element. This will include all the bacteria and fungus groups as well as smoke, decay and cooking odors. Ozone is not as effective on inorganic odors like ammonia, phosphates, nitrates, sulfates, chlorides, etc. The U.S. EPA states that there is not yet enough data available to determine exactly which chemicals ozone is effective against.(6) However, one odorous chemical compound which ozone has proven to be effective against is acrolein. Acrolein is also one of the many irritating chemicals found in secondhand tobacco smoke and it will break down when it comes in contact with ozone.(6)

As we pointed out earlier, the application of ozone has been successful in the restoration of homes or other buildings damaged by smoke. Smoke odor molecules that infiltrate all porous surfaces can be permanently removed by ozone gas. Only ozone will work on the most stubborn of all odor molecules, namely protein. Odors associated with decaying foods or animals, such as rodents, have long resisted normal chemical deodorizing attempts. Ozone has the potency to neutralize even these contaminants.

Ozone generators not only deodorize the surroundings by chemical oxidation but they also help to clean the air and may reduce the risk of microbiological infections. Ozone is a powerful, broad-spectrum antimicrobial agent that has been found to be effective against bacteria, fungi, viruses, protozoa, and bacterial and fungal spores.(7) The anti-microbial activity of ozone is based on its strong oxidizing effect, which causes damage to fatty acids in the cell membrane.

A big problem in infection control at health care facilities is that some strains of bacteria can actually *build up a resistance* to certain chemical disinfectants. Ozone, on the other hand, kills bacteria within a few seconds by a process known as cell lysing. Ozone molecularly ruptures the cellular membrane, disperses the cell's cytoplasm and makes reactivation impossible. Because of this, ***microorganisms cannot develop ozone resistant strains***; thus eliminating the need to change biocides periodically.(8)

Disinfecting means the use of a chemical procedure to eliminate virtually all recognized pathogenic microorganisms but not necessarily all microbial forms on inanimate objects. Antimicrobials such as iodine, chlorhexidine, 70% isopropyl alcohol solution, and hexachlorophene are frequently used in hospitals and other health care facilities. Chlorhexidine and hexachlorophene are active against many microorganisms but are less effective against Gram-negative bacteria. In general terms, *ozone has a much broader action.*

Applications of ozone are generally quick and easy as units are usually portable and only require a small amount of time to treat the average room. Ozone is not only an effective broad spectrum antimicrobial, but because it can be used in gaseous form, *ozone can give a complete coverage of all surfaces.* Certain ozone technology can also be applied with very little if any manpower requirement, once it has been installed.

What about clean rooms? They are used in many industries including food, beverage, pharmaceutical, research, analytical testing, and semiconductor production. Clean rooms are rooms which have had precautions put in place to make them free from possible biological contamination. Maintaining a high level clean room can be expensive and obtaining necessary funding to build these rooms can often be difficult.

Current methods for controlling the environment in clean rooms include HEPA filters, (which can be costly to maintain), and chemical cleaners (which are relatively inexpensive but present another set of issues). Clean rooms are designed for the purpose of reducing the particulate dust content and virtually all the air is recycled. Considering that you may have a dozen chemicals in the clean room at any time, the fumes and vapors are constantly entering and not being filtered out. Exposure to possibly toxic chemicals of persons working in the clean room is a real concern.(9)

Pharmacy compounding rooms are used by most hospitals to prepare medications and sterile intravenous admixtures for patients. The problem is that these rooms are often inadequate in their design to maintain a truly sterile environment.

Ozone technology has the potential for application in clean rooms and pharmacy compounding rooms as a disinfectant. It is a safer alternative than some current disinfectant chemicals. The low cost of ozone technology may also be appealing for industries which cannot afford some of the more expensive clean room technologies.

What about dust mites? - Will ozone help to get rid of them? The answer is a qualified Yes! Ozone does play a significant but indirect role in dust mite control. As we saw in Chapter 3, the ecological niche for dust mites is the consumption of skin flakes from humans and other animals. These skin flakes cannot be "fresh" but must be defatted. Skin flakes that serve as the food source for dust mites are made up of those defatted ones decomposed by the common mold known as *Aspergillus anastelodami*. Ozone application can help to control molds like this.

The effect of ozone in air seems so remarkable that one might wonder what is there that ozonation does not do. There's quite a bit. Ozone does not remove radon and other inert gases, chlorine, bromine, carbon dioxide, or the many small particulates in polluted air. Certainly, it does not by itself cure or relieve many of the adverse health consequences we described in Part One. But it does go a long way to help in cleaning the air and the result is a healthier environment for the occupants who share that space. Clearly, by improving indoor air quality, one's quality of life can only change for the better.

Setting Limits

The issue of safety should always be addressed when using ozone technology indoors with possible human exposure. Indeed, ozone air cleaners are a potential health risk if used at high levels in occupied indoor spaces. Current manufacturers of this type of device all seem well aware of this health risk and many continue to work feverishly to improve the science that will make ozone technology even safer for indoor use.

In the early days of this technology, all ozone generators used for air purification had the *potential* to produce ozone levels over the recommended safety levels. To combat this problem, some manufacturers of portable ozone air purifiers introduced models with ozone sensors that automatically shut off the units if ozone levels were to rise above a prescribed limit. This was a step in the right direction but the devices are not fool-proof, nor do they seem to work well in all size environments. They also do nothing to control the "blasting" phenomenon of high concentrations of ozone when in operation. This has been observed in earlier devices that tended to produce erratic, unpredictable spikes in concentration. That's hopefully a thing of the past, although not all air purifiers on the market comply. The U.S. EPA demonstrated over a decade ago that some devices on the market at that time were capable of producing ozone concentrations well above acceptable levels. For sure, a lot has changed since that time.

Ozone is

a powerful,

broad-spectrum

antimicrobial agent

Other manufacturers have built-in a safety design. The method used for producing ozone is based upon input voltage from a variable resistor (called a rheostat) to a transformer which amplifies the voltage. By this method, any component failure in either the control mechanism or transformer will result in less or no ozone production. In other words, if there are any failures in the ozone production process, it will occur on the side of safety, producing lower and not higher levels of ozone.

We will learn in the next chapter of the space-age developments in this field that ensure safe ozone levels. This unique space-age technology will not allow the accumulation of ozone to exceed recommended safe levels. It is a fail-safe method that does not rely on electronic gadgetry or inexpensive sensors that might fail. **There is no longer a reason to have serious concern about the potential harmful effects of too much ozone. That's no worry for the new space-age technology.**

That's enough said about this space-age technology for now. You must endure whatever suspense is in the making.

Finally, on the subject of Ozonation, lest you 'miss the forest for the trees', following is a summary of the major advantages of ozone technology that you can refer to later.

Ten Advantages of Ozone Technology

1. One of the biggest advantages of ozone may be its relatively **low cost** in comparison to other technologies.

2. Ozone will **neutralize virtually all organic odors**, specifically those that contain carbon as their base element.

3. Ozone is also **less corrosive to equipment** than most chemicals currently being used, such as chlorine.

4. Ozone generators clean the air and **may reduce the risk of microbiological infections.**

5. Applications of ozone are generally quick and easy as **units are portable** and only require a minimal amount of time to treat the average room.

6. Ozone has been found to be **an excellent disinfectant**, especially for the treatment of water and waste water.

7. Ozone is not only an effective broad-spectrum antimicrobial but because it is can be used in gaseous form it can give a **complete coverage of all surfaces.**

8. Ozone technology can be applied with very **little if any manpower.**

9. An important advantage of ozone use in food processing is that the product can still be called **organic**.

10. Ozone kills bacteria within a few seconds by a process known as cell lysing. Because of this, **microorganisms cannot develop ozone resistant strains,** thus eliminating the need to change biocides periodically.

With all those advantages, ozone technology has earned a rightful place as an air purification method. It is not the ultimate answer but it can make a useful difference if exploited appropriately. It is just the first of three conventional purification methods.

Now we move on to the second method: **Ionization**.

IONIZATION

As we observed earlier, many practical innovations in science and technology have come about from studying natural phenomena. Who could question the fact that nature does things best. Ozonation is just one example of how nature's methods of purifying the air have been copied for controlled use indoors. A second method that has been studied quite thoroughly is the natural phenomenon called ***Ionization.***

In simple terms, ionization is the process or result of a process whereby an electrically neutral atom or molecule acquires either a positive or negative electrical charge. Every high school chemistry student has learned that ionization occurs when 'energy in excess of the ionization potential is absorbed by an atom or molecule, yielding a free electron and a positive ion'. These free negative electrons or positive charges can then attach to other neutral molecules, particles or surfaces to make either of these acquire a negative or positive charge. The ions of the atmosphere have been of scientific interest for more than a century.

A Charged Atmosphere

You may be surprised to learn that there really is some (usually unintended) meaning to the phrase 'a charged atmosphere'. The air is truly charged - although not by mood, innuendo or sensitivity (as is often metaphorically presumed) - but by the sun's energy and the production of free electrons in space. Ionized air molecules make up only a very small percentage of the atmosphere but they play a most significant role. They can be quantitatively measured with an ionmeter using either a standard charged-plate collector or monitor, or an electrostatic-field meter that measures static decay on glass substrates. In numerical terms, on a typical fair weather day, ions in the air would number about 200-3000 ions/cm^3, of both polarities. Most ions increase during rainfall and thunderstorms: negative ions may increase to about 14,000 ions/cm^3, while positive ions only increase to about 7000 ions/cm^3. On the contrary, smoking one cigarette can reduce air ions in a room to about 10-100 ions/cm^3. The acceptable minimum of negative ions for healthy indoor air is on the low end of about 200-300 ions/cm^3, while an optimal level is about 1000-1500 negative ions/cm^3. However, typical values fall far short.

Why is that?

There are a number of valid reasons. Modern walls made of steel, or of iron and concrete, tend to create a Faraday Cage effect that shields the interior from outside ionization. Propelling air through metal ducts reduces negative ionization at a loss of about 20% for every two meters. It's worse with drier air and warmer ducts. Human activities introduce particular and chemical pollutants as well as microbes, which all take negative ions out of the air. In addition, all sources of fire (burning gas, woodstoves, fireplaces, smoking) directly add large quantities of positive ions that neutralize negative ions.

Ions help to maintain the health of the atmosphere by removing undesirable particulate and chemical pollutants. In the indoor environment, ionization provides an even wider spectrum of benefits including the destruction of bacteria and the elimination of odors. However, these indoor benefits of ionization tend to be negated by conventional construction and ventilation techniques, unless specific technology is employed to augment the ionization of indoor air.

Ionization devices

To induce ionization by separation of charge requires energy, which can be supplied either by natural process or by artificial means. Natural processes might include solar radiation and cosmic radiation, emissions from natural radioactive elements, electrical discharge (lightning), or frictional charging by wind and water droplet breakup (waterfalls, showers). Artificial sources could include strong electric fields (corona discharge), radio frequency generators and ultraviolet light.

Bipolar air ionizers create charged air molecules. When an electron is added or removed from air molecules, they are given a negative or positive charge respectively. There are three basic types of ionization sources that are currently available. They either derive their energy from photons, nuclear emissions or electrons. **Photon** ionization uses a soft x-ray energy source to displace electrons from the air molecules. **Nuclear** ionizers utilize polonium-210 radiation sources to create alpha particles (helium nuclei) that collide with the air molecules, displacing electrons. The molecules that lose electrons in either case become positive ions. Natural gas molecules rapidly capture the displaced electrons and become negative ions. These types of ion generators have no emitter points and create no concern regarding deposits. But, because of their high energy, they must be carefully installed and controlled to avoid creating safety hazards. They have found more limited application.

Electron ionization is the most common method in commercial use. It uses a corona-discharge ionizer with a high voltage applied to sharp emitter points or grids to produce a strong electric field. This field interacts with the electrons of adjacent gas molecules - extracting or adding electrons depending on the positive or negative applied voltage - to produce ions of the same polarity. Therefore, these ionizers can be classified by the type of electric current applied to the emitter, ie. pulsed DC, steady-state DC, or AC. The oldest type of electronic ionizer in common use employs alternating current (AC) so that the inherent voltage swings as the electric fields produced move from positive to negative. They produce alternate clouds of positive and negative ions with each cycle. These unipolar ions go on to form other chemical species depending on their ratio and concentration, the type of current, the mode of action and the relative humidity.

Historically, a fourth type of ionizer using *radio frequency fields* has been used and is still in use. But this remains an enigmatic application since radio frequency waves are on the low-energy end of the electromagnetic spectrum where non-ionizing radiation is the usual order of the day.

It is quite common to *label* or *classify* different types of devices that are designed as air ionizers, electrostatic precipitators and ozone generators in the same category. Although they all produce ions, directly or indirectly, they have distinct differences in modes of operation.

Air ionizers focus on the *emitter*, where the ions can be generated with some semblance of control to meet specific demands of a given indoor air environment. Particles become electrically charged through direct contact with the air ions.

Electrostatic precipitators are designed as much around the *collector* plates which become less effective as charged particles are deposited. Charged collector plates can attract even electrically neutral particles by induced polarity. They must be frequently cleaned.

Ozone generators are designed around the chemical changes associated with ozone, and its concentration must be rigorously limited in occupied spaces to avoid potential toxicity. Some energy processes that produce ozone also lead to ionization and vice versa. Since both consequences are directly involved in principle, in air purification, any device that produces either or both can be classified in the ozonation or ionizer category.

In all cases, the efficient production of ions provides a practical method for reducing microbial contamination and neutralizing odors, by destroying and/or removing volatile and particulate constituents from the indoor air.

There is quite a long history to the various applications of air ionization technology. Most recently, it has been important in the control of electrostatic discharges (by neutralization with air ions) in the manufacture of sensitive materials like semiconductors. With the development of efficient automated controls, ionization is being increasingly applied for air cleaning where more stringent controls are required.

There are two side effects of this ionization technology that have somewhat limited its application. First, the ionization process is often accompanied by *ozone* formation which, if not controlled, can lead to problems. The second effect is that of *soiling* whereby, as we pointed out before, the charged particles become deposited on walls, furnishings, carpets etc. and become a nuisance.

One patented approach to reducing these problems has been popularized by a small California company. The practical solution called cross-field gas ionization employs electric fields to generate a low energy stream of electrons, while a steady flow of air passes perpendicular to the orientation of that field. Some of the air molecules pick up electrons as they are passing across the electric field, thus becoming negatively charged. This innovation is now in popular use in some supermarkets, typically located close to the fish counters.

Negative Ionization

All sources of ionization have the common effect of electrifying the atmosphere. At ground level, the ratio of positive to negative air ions is normally about 1.1-1.3, decreasing to about 0.9 following certain weather events. However, most research to date suggests that the negative ions are more favorable for clean, healthy air quality and an environment conducive to human health and well being.

The major constituents of air are nitrogen and oxygen. When air ionization occurs, it should be no surprise that these are the primary players. There is quite a complex set of species generated and a plethora of subsequent chemical reactions,(10) too detailed to be elaborated on here. But it is important to underline the key role of some specific reactive oxygenated species at least.

When electrons are removed from the normal air gases, nitrogen and oxygen, the initial primary positive ions are N_2^+, O_2^+, N^+ and O^+, all of which are very rapidly converted (in microseconds) to different protonated hydrates. That is, they extract hydrogen nuclei H+ from water and become associated with small groups of water molecules (less than 10 per ion). The free electrons quickly attach to oxygen to form the superoxide radical anion $O_2^{\cdot-}$ which can also form hydrates. These are all called 'cluster ions'.

The small primary ions and the cluster ions have ample opportunity to collide and react with any and all constituents in the atmosphere. Remember this is all on a microscopic scale. It is a fair estimate that at ground level, a typical cluster ion might live for about one or two minutes, and during that time, make as many as 1,000,000,000,000 (10^{12}) collisions with other species.(10) This is a dynamic interactive environment, full of charge and energy. It is invisible to the naked eye, but no less real.

In this interactive process, superoxide is just one of a group of negative ions that are derived from oxygen. These include oxygen, superoxide, peroxide, hydro-peroxide and hydroxyl species that are altogether termed **'reactive oxygenated species' (ROS)**.

These reactive oxygenated species can participate in a number of oxidation-reduction reactions in both gaseous and aqueous phases. In the atmosphere, they are very significant in the destruction of volatile organic chemicals, the removal of particulates, the formation of smog and the conversion of ozone. We will return to this subject later in the space-age technology to be described in the next chapter.

As negative ions move through the air, they first attack, then become attached to and finally, convey their electric charge to particles of dust, smoke, water and their associated allergens, VOC's and other contaminants. These 'particle clumps' are clearly heavier than air so they can no longer float around but tend to fall to the ground for vacuum removal. This is a ***physical effect.***

These same oxygenated species can also have a ***chemical effect*** since they are all very reactive. The reactions with organics especially are again quite complex. But the net result is usually quite simple. Organic chemicals, including hydrocarbons and their derivatives specifically, are reduced by these ROS to the lowest common denominators of water and carbon dioxide.

Just to illustrate, consider the effect of ROS negative ions on formaldehyde (CH_20). It first forms water (H_2O) and carbon monoxide (CO) which becomes further oxidized into carbon dioxide (CO_2). These are harmless, odorless products. It is therefore not surprising then that ionization (which also produces ROS) also helps to eliminate many odors which are so often associated with these organic contaminants.

Efficiency of Ionizers

Negative air ionization also has the potential to reduce the concentration of airborne microorganisms. Such a ***biological effect*** appears to result from two phenomena: the ionization of bioaerosols and any dust particles that may carry microbes, causing them to settle out more rapidly (comparable to the physical effect above) and also, the possible reduction of microbial viability by inactivating viable micro-organisms that remain airborne.

In effect, therefore, the overall air purifying efficiency of this ionization process (technology) is the product of all three: the physical, chemical and biological (bactericidal) efficiencies of any ionic air device.

Several studies have been published which illustrate the wide variety of applications to which this ionization technology has been successfully applied to reduce microbial air pollution. For example, positive results have been reported for the reduction of the incidence of Newcastle Disease Virus in poultry houses;[(11)] the reduction of colony forming units in a dental clinic;[(12)] the reduction of *S. aureus* bacterial aerosols in patient rooms of a burns and plastic surgery unit,[(13)] and reduction in levels of aerosolized virus T1 bacteriophage recovery.[(14)] Those are just a few applications.

Perhaps the best peer-reviewed research on ionic purifiers has come from Professor Sergey A. Grinshpun and his colleagues at the Center for Health-Related Aerosol Studies, Department of Environmental Health, University of Cincinnati, Ohio.(15-17) Over the past few years they have been able to demonstrate unequivocally that ionic purifiers efficiently *remove* fine and ultrafine particles from indoor air in a typical room. The particle removal efficiency reached almost 100% in a walk-in test chamber. They significantly *reduce* the aerosol concentrations (down to virus size) in the breathing zone of a test manikin. They show the ability to *cause* germ inactivation and to greatly *enhance* the effectiveness of different types of respiratory masks in protecting against virus-size aerosol particles. As such, they should make a crucial difference with respect to exposure and health risk against airborne biological agents. There will be much more on this later, in terms of application.

The U.S. Department of Agriculture has reported recent studies on the effectiveness of ionizers.(18)

- Ionizing a room led to 52% less dust in the air and 95% less bacteria.
- Negative ionization removed airborne *Salmonella Enteritidis*: "These results indicate that negative air ionization can have a significant impact on the airborne microbial load in a poultry house and at least a portion of this effect is through direct killing of the organisms."
- The use of an ionizer resulted in dust removal efficiencies that averaged between 81.1 and 92.2% in a poultry hatchery.

Conceptually, ions in the air can have at least four different fates: (i) they can recombine with other ions; (ii) they can react with gaseous molecules; (iii) they can become attached to larger particles, and (iv) they can make contact with a surface. In principle, the first two of these processes would be involved in the removal of volatile organic compounds (VOCs), while the latter two would account for the removal of particulate matter (PM).

The bottom line is that **air ionization technology has been proven effective as a method of air purification.** Physically, it removes particle contaminants; chemically, it helps to neutralize VOC's and odors, and biologically, it removes bioaerosols and particles carrying microbes. There is therefore no doubt that this technology can appropriately address many issues related to indoor air quality.

Some major advantages of ionization technology are summarized for your convenience on the opposite page.

Ten Advantages of Ionization

1. Efficient in **removing fine and ultra fine particles** from indoor air.
2. Can be effective in the **submicron range** of bioaerosols particularly.
3. Effective in removal of hazardous **VOC's**.
4. **Enhances routine respiratory protection** of surgical masks and dustmist respirators.
5. May have a special role in **defending against bioterrorism**.
6. Produces less hazardous reactants and **stable byproducts**.
7. Generates minimal bulk deposition of particles (**no soiling**).
8. Results in extended and **improved performance** of other Conventional Technologies, like filtration and adsorption.
9. Has **Low Energy Costs.**
10. Has **Low Maintenance** Requirements.

In addition to **Ozonation** which we described earlier, **Ionization** has proven to be a second effective Method of the Air Purification strategy which we identified in our search for **Fresh Air for Life** as **Strategy #4**. We now go on to address the third Purification Method: **Ultraviolet Germicidal Irradiation.**

IRRADIATION

Ultraviolet Germicidal Irradiation (UVGI) was first investigated for air disinfection back in the 1920s by William F. Wells at the Harvard School of Public Health, soon after he described the 'droplet neuclei' theory of airborne disease transmission.[19] That work was later advanced by Dr. Richard Riley and others at Johns Hopkins Medical School, with a focus on tuberculosis control. Initial results were quite promising but then, with the development of anti-tubercolosis drugs and a later vaccine, the UVGI method was never fully established.

The history of UVGI and air quality has been one of varying success and inconsistent performances. This technology has indeed had its ups and downs. At times it showed great promise as a method of controlling the spread of infectious disease, particularly when airborne epidemics were threatening. However, it remained unpredictable. The design of UVGI air disinfection systems was then an art in search of a science.(20) Until recently, there has been a lack of focused research and analysis. But times are changing.

UVGI has been used in some healthcare facilities (60%), prisons (19%), shelters (19%), and schools (< 2%) for example, but only in recent times has there been a keen revival of interest in this technology. There has been only limited use in domestic households.

This new focus on UVGI has been driven by a resurgence of tuberculosis in the United States since 1985 (though followed by a subsequent decline), as well as growing concern over the spread of disease, including multi-drug resistant strains, in different parts of the world. Widespread international travel heightens that concern, as well as the potential threat of bioterrorism (including multi-drug resistant TB) in the United States. All defensive and protective public health approaches must now be reconsidered.

UVGI Technology

UVGI technology is based on the simple but profound observation that microorganisms are uniquely vulnerable to the effects of light at wavelengths at or near 254nm due to the resonance of this wavelength with molecular bonding and structures. In other words, the energy of this particular light is just about right to break organic molecular bonds apart, which translates into cellular or genetic damage for microorganisms. They simply become impotent (cannot reproduce, and therefore become non-infectious) or just plain *die* as a result of this damage. All viruses and almost all bacteria (excluding spores) are vulnerable to moderate levels of such UVGI exposure.

Ultraviolet light covers the energy region of the electromagnetic spectrum just above the tail of visible 'violet' light. Since so much is said in the media about UV light, we should identify more precisely which of the different bands of UV is important here.

UV-A refers to the near or long wave length (**320-400nm**) UV band (sometimes referred to as 'black light'). It makes certain pigments fluoresce but for our practical purposes it has little effect on microorganisms and no known effect on human tissue.

UV-B covers the middle UV band (**280-320nm**) and is commonly used in tanning salons (we think, unfortunately so!). It does inactivate some microbes and we know that prolonged exposure can cause blistering of the skin, damage to the cornea of the eye and contribute to skin cancer.

UV-C is the germicidal or short wave length **(200-280nm)** UV band that is of particular interest here. It is actually referred to as ultraviolet germicidal radiation (or UVGR). This will inactivate microorganisms on exposure by destroying their DNA, RNA and vital proteins, while having little effect on humans.

We should also note that there is a higher energy segment of the ultraviolet region in the electromagnetic spectrum - called the **UV-V** or UV-Vacuum (**below 200nm**) region. This includes the 185nm wavelength, which we will see later (in the following chapter) has specific application in the important photocatalytic excitation process.

Germicidal irradiation technology just happens to benefit from a number of fortuitous properties of both UV-C and the droplet nuclei that it targets:

- The optimum wavelength for destroying biological effects is close to 254nm which just happens to be a narrow band generated by common inexpensive mercury lamps (how **economical!**).
- The short wavelength UVGR, although inherently more biologically active, is much less penetrating into matter (how **safe!**).
- The minute size of airborne particles renders them highly vulnerable to UVGR, despite its limited penetrating capacity (how **effective!**).

This is the Super Intelligent Design that we can copy. What nature does outdoors, we can arrange to reproduce indoors. When demonstrated under idealized conditions in the laboratory, UVGI achieves extremely high rates of microbe mortality. But in actual real world applications, other factors reduce the effectiveness. These factors include:

- The exposure time (air velocity must allow for a sufficient dose).
- Mixing of room air (for non-powered 'upper room' applications).
- Power levels (effects will vary with UV intensity).
- Moisture (humidity) and particulates (both 'protect' microbes).
- Dust on lamps (reduces exposure - therefore, regular maintenance).

Despite these considerations, UVGI is still a proven effective method for disinfecting air. A recent six year study by the University of Colorado for the CDC's National Institute for Occupational Safety and Health (NIOSH) has demonstrated that, when specified correctly, UVGI is extremely effective in reducing or eliminating airborne pathogens.(21) The study used clinical data, noting the reduction of pathogen colonies, to quantify the results of their testing in a standard hospital room.

In the March-April 2003 issue, *Public Health Reports* examined the use of UVGI as a bioterrorism counter measure with extremely positive results. The article was entitled '*The Application of Ultraviolet Germicidal Irradiation to Control Transmission of Airborne Disease: Bioterrorism Countermeasure*'.(22) Weaponized pathogens are just as susceptible to the effects of UVGI as are the pathogens encountered on a daily basis in healthcare facilities. They recommend a multiple UVGI attack on pathogens. This can be achieved by installing UVGI devices in the 'upper room' area, the ducting system and the heating/cooling coils in the air-handling units. That matter is still reserved for the professionals at this time.

UVGI Applications

As a method for air purification (and disinfection, especially) UVGI has its place. But it is not by any means a substitute for any of the other methods. It is clearly not a replacement for HEPA filters which remove many small bioaerosols and fine particles that carry microorganisms. The different methods and strategies complement each other and the space-age technology we will describe in the next chapter will exploit the synergy among them all, and then crown the approach with a surprising technical innovation.

UVGI is still a proven effective method for disinfecting air.

But where should UVGI be installed anyway?

The most recent data available on the types of UVGI systems that are currently being installed indicates as follows: Microbial Growth Control (31%), In-duct Systems (27%), Upper-Room Air (25%), and Room Recirculation (17%).(23) Let's look at all four of these in turn.

Microbial Growth Control. In Europe, microbial growth control on cooling coils has been practiced in breweries since at least 1985, especially because the wrong fungus can cause big spoilage problems. In North America, the focus has been on controlling growth on central heating/ventilation systems. It is well known that microbes tend to collect in the heating and cooling coils of HVAC systems which are inherently subject to moisture or high humidity. Germicidal lamps are rapidly becoming popular as an easy fix for the air conditioner coil which is prone to have a mold problem. The unirradiated coils show natural contamination from various fungal species, including *Aspergillus* and *Penicilliun*. Bacteria and even algae are not uncommon. Viruses are intracellular parasites and therefore would not normally replicate there. This microbial growth is the most prevalent cause of the 'mildew odor' you sometimes get, when you enter an air-conditioned sick building.

Low cost UVGR (254nm) lamps are easily installed near to the coil surface and have proven to be reasonably effective at suppressing mold growth on the coil. The effectiveness may be due to inactivation of the spores or perhaps because mycelial growth cannot be sustained under continuous exposure. The lamps are basically similar to sunlamps and are typically only effective on microbials that come within a few inches of the lamp, or areas where the light shines directly for extended periods of time. To cover most of the coil, several lamps must be used. The system is not effective at killing airborne mold unless numerous lamps are used.

In-Duct Systems. Installing an in-duct system at a point just before entry into, and just after the air exhaust from the room, will lessen the threat of airborne pathogens from either entering or leaving the room and destroy any that may be resident within the room. The HVAC ducting in any facility will not be 100% secure from leaks into and out of the ducts, so installing these UVGI systems is a prudent measure in preventing secondary infections.

Since UV rays are primarily emitted perpendicular to the surface of the lamp, the UVGI lamps should be located at right angles to the airflow, so that the rays are emitted parallel to the airflow. This will maximize the exposure time for airborne bacteria as they flow through the HVAC system. Placing a lamp on the duct wall parallel to the airflow is highly inefficient. The size of the lamp fixture required will depend on the duct or plenum size, the length of the duct compartment where the unit is to be installed, the typical air speed and the approximate air temperature. Mounting the fixture usually requires cutting a hole within the duct wall to insert the lamps across the duct and attaching the lamps to the wall of the duct with brackets. Continuous use is recommended since cycling germicidal lamps may reduce their rated life disproportionately.

Upper-Room UVGI. This is an interesting design feature that combines efficacy and safety in a local environment. Fixtures are mounted in the ceiling or near the ceiling on adjacent walls and from there, they irradiate and disinfect the air in a large portion of the upper room, above the heads of occupants. This is a really cool design. Here's what happens: Air slowly rises from the lower room into the irradiated zone, driven by body heat and other heat sources, body motion, drafts, and forced air from fans or HVAC systems. The UVGI destroys the bacteria and viruses that are carried into the UV field by the convection currents of air circulation.(19) Disinfected air from the upper room then descends and dilutes infectious particles in the lower room. The process is silent, essentially draft-less and energy efficient, but it takes time to be effective.

This is useful for retrofitting older buildings with or without HVAC systems. It is particularly suited for large spaces such as lobbies and waiting rooms, areas that are difficult to adequately treat with high rates of either ventilation or filtration.

Dosing guidelines date back to the 1970's but are still useful. Approximately one 30 Watt suspended fixture, or two 30-Watt wall fixtures, should be used for each 19 square meters of floor space. It is important to monitor the UV radiation in the lower occupied level and to maintain it below acceptable limits. Most UVGI overexposure incidents arise from accidental direct exposures of workers to germicidal lamps in the upper room where fixtures were not turned off, or where installations of fixture designs were clearly unsafe. The American Conference of Governmental Industrial Hygienists set a standard of 0.2 microwatts/cm^2 for 8-hour exposure. Easy-to-use handheld meters are readily available to check the intensity of aging UV lamps and to survey installation areas for exposure levels.

Room recirculation. Self-contained UVGI exposure chambers are used in residential, commercial and industrial applications. Fixtures are commercially available in a number of different configurations to adapt to virtually any setting. They can be mounted on a ceiling or wall, or made available for portable or mobile use. Air is drawn into the fixture by a fan, through a washable electrostatic particulate filter. It then is forced into an ultraviolet exposure chamber where it is irradiated by UVGR. Purified air leaves the exposure chamber through the louvered exhaust panel. Obviously, fixtures can be sized appropriately, depending on the dimensions of the room and air changes per hour required by the application. Although the effect of re-circulating the room air is partially dependent on air-change rate, the result is an effective increase in removal rate of microbes in comparison to a single-pass system.

There are some advantages to this type of design. The fixtures can be tailored to most effectively improve room air quality. There is no installation cost per se - just plug in, turn on and *voila*! -- So easy to use. There is no real risk of accidental UV exposure of the occupants in the treated area, and perhaps best of all, it is relatively portable so that it can be applied wherever, whenever and for as long as necessary to get the best job done.

The Big Picture

UVGI systems are often used in combination with HEPA filters. This practice is usually recommended for applications in isolated rooms. For many other applications, however, the advantage of HEPA filters over a regular high efficiency filter in removing microbes does not warrant their exclusive use with UVGI.

In all these systems, UVGI suffers from the same limitations as we found before, when using air filtration or ventilation as a strategy of disinfection. Air can still be moved through the device at a rate sufficient to double or quadruple the existing ventilation rate, only to reduce the risk of microbial infection by just a half or three-quarters. That would definitely be inadequate in almost every infectious situation. Plus, attendant noise and drafts from high rates of air movement can become objectionable to room occupants. But it is still helpful to conclude that **UVGI is economical, clean, relatively safe and offers little or no resistance to airflow**.

Again for completeness and for your convenience, the advantages of this UV irradiation method are listed here.

TEN ADVANTAGES OF IRRADIATION TECHNOLOGY

1. **Low cost**
2. **Easy installation**
3. Effective against **molds**, eg. AC coils
4. Effective against many **bacteria and viruses**
5. Useful in **healthcare facilities**
6. Effective in **congregational settings** (Upper Room UVGI)
7. **Negligible airflow resistance**
8. Can meet federal ozone **safety guidelines**
9. Safe - **Low penetrating UV-C** on exposure
10. **Low maintenance**

Speaking of economics, it might be interesting to note that in commercial offices, the cost of UVGI installation could in the long run prove to be cost-effective, compared to yearly losses from absenteeism because of building-related illness. A recent year long study of 771 employees from three different office buildings in Montreal, Canada showed that the UVGI technique reduced overall worker sickness by about 20%, including a 40% drop in breathing problems.(24) The authors estimate that the installation of UVGI in most North

American offices could resolve work-related symptoms for about 4 million employees. Dr. Wladyslaw Jan Kowalski, an architectural engineer at Penn State University's Indoor Environment Center went even further: "*Theoretically, if a large number of schools, office buildings, and residences were modified, a number of airborne respiratory diseases could be irradicated by interrupting the transmission cycle. Reducing the transmission rate sufficiently would ... halt epidemics in their path*".[25] That is indeed a bold projection, but could it have some reasonable foundation or justification? Just perhaps.

Such optimism might appear to the uninformed observer as unjustified exaggeration and a far cry from reality. But it would only seem that way until you learn about the space-age technology and its application to indoor air pollution which will be described in the following chapter.

So now, for a *journey into imagination* -

* **If you could take all the air purification technologies that have been described up to now and combine them into a single efficient system ...**

* **If you could exploit the best state-of-the-art understanding of photo-oxidation and atmospheric chemistry ...**

* **If you could find a solid state catalytic system to generate the right oxidative species to disinfect, deodorize and detoxify the air ... and finally ...**

* **If you could do this in a safe, simple and cost-effective way for widespread application ...**

Then, and perhaps only then, any increasing hope to win the unseen battle against indoor air pollution would certainly be justified. You might just then gain access to *Fresh Air for Life*.

That's what we will discover next. Your wait is over!

STRATEGY #5

TEN

Use Space Age Technology

"It took the synergy of ***Five Essential Features*** *to deliver* ***the Ultimate Solution!"***

Space technology touches and enhances practically every aspect of life. In fact, much of the technology that improves our lives so much today, benefits directly from some technology that was originally developed or adapted for space exploration. Such are the products and services that set the standard for comfort, convenience and reliability.

When President Kennedy in 1961 committed the United States to placing '*a man on the moon before the end of this decade*', he unleashed a force of creativity and innovation that spawned a revolution. When Neil Armstrong in the summer of 1969 took his famous '*one small step for man, one giant leap for mankind*', he took a step into a new generation of ideas and technology that was destined to change the world. And that it has.

Air purification was not left out and just as well.

INSIDE A SPACE CAPSULE

All the astronauts and cosmonauts who have inhabited those small space capsules that temporarily became their homes, have had to come to terms with indoor air quality as they experienced it. Living in those cramped quarters for days on end was indeed a challenge in many areas, but not the least of these, was the problem of finding fresh air to breathe. Or even worse, consider the brave pioneers who have occupied the international space station (ISS) for months at a time. They could never 'open the window to let in some fresh air'. We've all seen them on television, breathing ambient air inside their adopted homes, without any masks or spacesuits. But as you would well imagine, having three to seven people sharing a small enclosed volume on the space station for weeks and months at a time, poses a critical air management problem for space scientists and engineers. The astronauts have to contend with air as they make it … contaminated or purified, as the case may be.

The truth is that even the cleanest of human environments is full of microbes. **Table 11** lists the variety of airborne microorganisms that have been isolated on American and Soviet spacecraft.(1)

Table 11 Airborne Microorganisms Isolated in Spacecraft

Bacteria	Apollo	Soviet	Shuttle	OA*
Acinetobacter	Y	1	Y	
Staphylococcus	Y	3	Y	
Corynebacterium		7	Y	
Streptococcus		6	Y	
Moraxella	Y	2		
Klebsiella pneumoniae	Y	4		
Haemophilus influenzae	1			
Haemophilus parainfluenzae	1			
Mycoplasma	1			
Alkaligenes		1		
Neisseria spp.		5		
Pseudomonas spp.		5		
Bacillus spp.		9	Y	
Fungi				
Aspergillus		11	Y	Y
Cladosporium		2	Y	Y
Penicillium		13	Y	Y
Mucor		1		Y
Alternaria			Y	Y

* OA - Outside Air

There probably are other contaminants that pervade the air in every space capsule, especially after days or weeks in orbit. They most likely fall into the same broad categories: some fine **particles** originating from 'who knows where'; plus volatile organics, including **chemicals** from some experiments and materials degassing just like they do on *terra firma*, in addition to those same ubiquitous **microbes** that contaminate everywhere when given the chance.

Essential oxygen in space is made available by the process called electrolysis. Electricity from solar panels is used to split water into hydrogen and oxygen. On earth, this is achieved by the photosynthesis of plants as they take carbon dioxide and water, in the presence of sunlight, to make sugars for food and release oxygen into the atmosphere. To ensure crew safety in space, there are always large tanks of compressed oxygen in reserve.

In addition to exhaled carbon dioxide, normal people (astronauts included) also produce and release small amounts of other gases. Methane and carbon dioxide are produced in the intestines and sometimes released. Ammonia is created by the breakdown of urea in sweat. Then there are by-products of metabolism - like acetone, methyl alcohol and carbon monoxide - released in urine and exhaled in air. 'Outgassing' from a crew mate is not unthinkable. All these gases (and others) will tend to accumulate over time, when the astronauts must of necessity be confined to their limited quarters on a Space Shuttle or in a Space Station, especially for extended missions. This is a real potential hazard. How **stifling!**

To efficiently remove all these chemicals from the air, activated charcoal filters are just a start. In February 2002, four astronauts aboard the *ISS Expedition* had to confine themselves temporarily to the Russian segment of the space station after detecting a foul odor in the U.S.-built Quest airlock. The offending but non-toxic odor originated in a recyclable metal oxide (Metox) canister used to clean and recharge air scrubbers for U.S. space suits. They had gone to descrub some filters used in the air scrubbers and in the process, released some noxious compounds. How **surprising!**

In 1997, a fire aboard the Russian space station *Mir* also gave NASA researchers samples of contaminated air. They were collected by astronaut Jerry Linengar after a faulty oxygen-generating canister sent flames and smoke billowing through the orbital outpost. The crew also had to deal with rising carbon dioxide levels and leaking antifreeze fumes. How **scary!**

Those incidents underscore what *can* happen on any given day (or moment) in space.(2) They point to a common problem that NASA must deal with, whether or not accidents ever occur.

NASA FACES A PROBLEM

Scientists at the National Aeronautics and Space Administration (NASA) have been fully aware of this problem of indoor air quality and have sponsored research in the field. It has been one of NASA's priority missions for decades now, to keep astronauts healthy in space where there are no alternate sources for fresh air, clean water or healthy food, should a craft's supply ever become contaminated.

With the real threats of biohazards and bioterrorism facing the nation especially since 9/11, the Department of Homeland Security has turned to NASA (along with other federal agencies) to draw upon its long experience of monitoring and preserving air, water and food in a closed indoor environment. In August 2002, the NASA Administrator appointed a senior advisor for homeland security who led the charge to identify technologies that were not only useful to the space mission but also to homeland security. Many of these dual-use technologies became projects under the purview of the Office of Biological and Physical Research (OBPR) which has funded research for a long time into ensuring air, water and food safety for astronauts.(3)

One of these dual-use technologies, originally devised to meet a problem in space, turns out to be at the center of the development of air purification technology for application right here on *earth*.

What was NASA's Problem?

Growing green plants in microgravity as a possible source of both essential oxygen and vegetable crops for astronauts on long-duration missions is a significant development program for NASA. But conducting this plant research in microgravity aboard the Space Shuttle or the International Space Station (ISS) requires a research facility or chamber that is enclosed and environmentally controlled. However, plant physiologists have known for a long time that growing plants produce ethylene - an odorless gas that acts like a hormone to plants and causes them to mature early and go to seed. (It is this same ethylene gas that causes bananas to ripen in a few hours when kept enclosed in a paper bag). In fact, concentrations of ethylene as low as 50 ppb (parts per billion) will affect plant reproduction. Concentrations of 200ppb may interfere with the pollen process. On earth, the ethylene normally diffuses into the atmosphere and is dispersed uneventfully. However, in the enclosed growth chambers in space, the ethylene would accumulate to levels that prove lethal to plants within days. The challenge was to find a way to 'scrub' the air and remove or degrade the ethylene efficiently.

For very limited space shuttle flights, NASA had used filter paper impregnated with chemicals that absorb ethylene. But that involves consumable absorbents that must be continuously replenished. Therefore, the goal was to develop a non-consumable degradation technology for scrubbing the air.

NASA Finds a Solution

NASA found the solution they were looking for in the early '90s. An ethylene scrubber was originally devised by Dr. Weijia Zhou, Director of the Wisconsin Center for Space Automation and Robotics (WCSAR) at the University of Wisconsin.(3,4) It was based on some unique properties of the white pigment, titanium dioxide (TiO_2). **When exposed to ultraviolet light, the TiO_2 acts as a photocatalyst** and oxidizes ethylene and other hydrocarbons into carbon dioxide and water vapor. In that way, you get rid of the ethylene, form totally harmless by- products and leave the catalyst unchanged. That's a perfect NASA solution!

This result has proven to be so useful and effective, that over the past decade WCSAR plant growth chambers equipped with photocatalytic ethylene scrubbers have been flown on nine Space Shuttle Flights, three ISS missions and even once on the Russian space station, Mir.(3)

But that's not all. In 1998 John Hayman Jr., Chairman of KES Science and Technology Inc. in Kennesaw, Georgia, licensed WCSAR's technology to develop a commercial ethylene scrubber for florists and grocers to extend the shelf life of flowers, fruits and vegetables kept in an enclosed cold storage system. He called it Bio-KES. Then with the first anthrax attacks late in 2001, one of his managers wondered out loud at a managers' meeting if the Bio-KES unit could possibly kill anthrax spores.

The first test with the basic unmodified unit killed 83% of *Bacillus thuringiensis* which consists of spores similar to the original *Bacillus anthracis*. Then they replaced the Bio-KES germicidal bulb with 52 UV lamps for high-energy exposure and found the kill rate was raised to 99.99998% of spores. That was deadly. Spores have hard shells, so killing the more normal live (vegetative) bacteria or even viruses could be a breeze. A new weapon against bioterrorism had been devised and by 2003, the U.S. FDA had given clearance for it as a class II medical device (for use in hospitals).(3) That's called serendipity. What would benefit the astronauts over time could also improve the quality of life on earth here and now.

Now, for the main purpose that we are concerned with here …

The ethylene scrubber technology led to the development of an air purification device which is effective against microbes. In the early applications, a fan would draw in room air and force it through a maze of tubes. As the infected air passed through the tubes, chemically reactive hydroxyl and possibly other radicals formed at the surface of TiO_2 when exposed to the high-energy UV light. These radicals could then attack and kill the airborne pathogens.

That was the beginning of photocatalytic oxidation (PCO) and its application to air purification … just the beginning!

PHOTO-CATALYTIC OXIDATION (PCO)

Back in the early 1970's, Japanese researchers discovered that water cleavage could be induced by the action of ultraviolet light on titanium dioxide electrodes(5). Since that time, photocatalysis with TiO_2 has attracted attention as an alternative technology to aid in the purification of both water and air.(6-8)

Conventional Photo-Catalytic Oxidation (PCO) air purifiers draw air into a photocatalytic chamber to be purified. The photocatalytic chamber consists of two critical elements: a high intensity UV-C ultraviolet bulb (typically radiating at 254 nm wavelength) and a titanium dioxide catalyst. When exposed to the ultraviolet light, the photo-catalyst becomes highly reactive and attacks the chemical bonds of the bioaerosol pollutants, thereby converting the toxic compounds into benign end products, namely carbon dioxide (CO_2) and water (H_2O). **In a nutshell, the UV light excites the catalyst to produce reactive species that oxidize or neutralize the pollution.**

Figure 2 illustrates the essential features of PCO. To grasp a clearer understanding of how this process works, we need to look at four essential components:

(i) The metal catalyst itself.
(ii) The excitation process.
(iii) The conversion of pollutants.
(iv) The germicidal effects of the ultraviolet light.

(i) *The Metal Catalyst*

Conventional PCO has exploited some important characteristic properties of the titanium metal. Titanium itself has been described by different adjectives that point to these properties: namely, it is light, strong, anticorrosive and relatively cheap. But it is the anticorrosive quality that is in play here. Titanium is readily oxidized and forms a unique, very thin layer of oxidized film on its surface that is the titanium dioxide (TiO_2). This we know as the key photocatalyst. Titanium dioxide itself is a multifaceted compound that has different uses commercially. It's the same stuff that makes toothpaste white and base paint opaque. But as a surface layer here, it is also a potent photocatalyst that can break down almost any organic compound.

The invisible TiO_2 barrier is quite remarkable indeed. If it is ever scratched or damaged, it will immediately restore itself in the presence of air or water. In that regard, titanium is similar to its chemical cousin: aluminum.

Throughout the PCO process, the TiO_2 is not consumed, but it remains as an auto-cleaning, reusable catalyst. **How convenient!**

Fig. 2 PHOTOCATALYTIC OXIDATION (*1st Generation*)

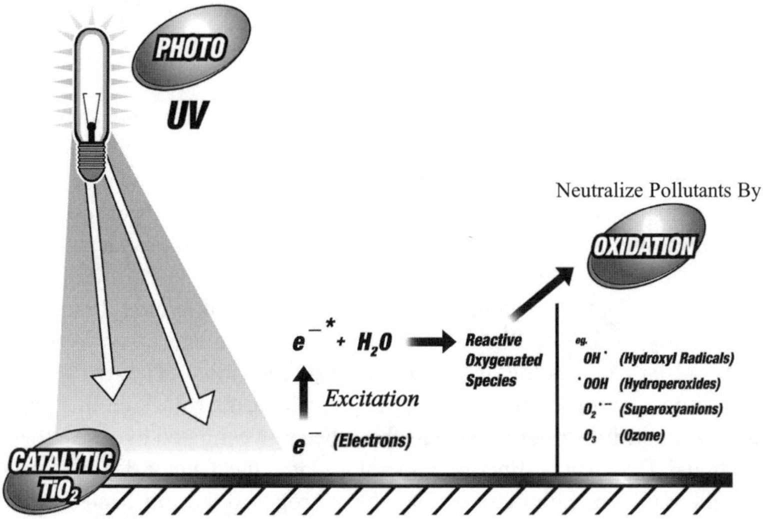

(ii) ***The Excitation Process***

When the ultraviolet light strikes the surface of this TiO_2 photocatalyst, it causes the excitation of surface electrons. These electrons are normally held on to the surface molecules but if given enough energy, they can overcome that restraining electrostatic force and become free. The threshold energy is defined as the 'band gap' and is a measure of the difference in energy between what are called 'valence' (or attached) electrons held by molecules on the surface, and the potential 'conducting' (or free) electrons that are not bound on to the surface. For TiO_2, the band gap energy is 3.2eV, equivalent to photons with wavelengths about 385nm or less. This is the so-called photoelectric effect which was discovered by the most famous scientist of the last century. You guessed it … Albert Einstein. That's what earned him his first Nobel prize back in 1921. That was before he proposed the Theory of Relativity and all his other theoretical ideas that also changed the world.

When this excitation phenomenon happens, the energy from the UV light is converted into electricity. What in effect is created is a set of electron/hole pairs that have sufficient lifetimes in the space-charge region to enable them to participate in chemical reactions. The holes are positive charges that, in the presence of water vapor, are believed to attach themselves to hydroxyl groups (anions, OH^-) to form free hydroxyl radicals ($^{\bullet}OH$), while the electrons might attach to oxygen molecules to form superoxide ions ($O_2^{\cdot -}$). Recall that both hydroxyl radicals and superoxide ions are highly reactive species. **How exciting!**

(iii) ***The Conversion of Pollutants***

The charges on the catalytic surface attract pollutant molecules, including hydrocarbons and other organics, which are readily attacked by the adjacent hydroxyl radicals and the other reactive oxidation species. The latter are quite unstable and simply need to attach to something or other. The net result is that the pollutants are oxidized in the process to the benign end products: namely, carbon dioxide and water.

One cannot overestimate the oxidative power of hydroxyl radicals. It is second only to that of fluorine and surpasses ozone and other oxygenated species, and far exceeds the oxidizing power of chlorine. They will also kill and decompose adsorbed bioaerosols. The mechanisms by which this process of photocatalysis kills microorganisms have been investigated by researchers at the National Renewable Energy Laboratory and the Center for Indoor Air Research in Golden, Colorado.(9) They have proposed that for bacteria like *E. Coli,* the reactive oxidation species first attack the cell wall of the microorganisms to increase their permeability. This is followed by cytoplasmic membrane damage leading to a direct intracellular attack and bacterial death.

That kind of stuff is lethal. It has been estimated that the energy field in the photocatalytic chamber is four orders of magnitude (x 10,000 times) more powerful than natural sunlight. That should give you some idea of the power of this photocatalytic oxidation (PCO) technology. **How powerful!**

(iv) ***Germicidal Effects***

Finally, we should also observe that the UV-C light (at 254nm) has the same germicidal action. It kills microorganisms including mold and bacteria. As we saw earlier, UVGI is applied effectively for this purpose at this wavelength. It inactivates microbes by disrupting their DNA and RNA to effectively sterilize the air. It is very effective when properly applied. **How effective!**

This simple conventional PCO technology has been in use for quite some time now, and has many attractive and practical features. Here's a short list.

Ten Practical Features of Conventional PCO

- High efficiency at room temperatures
- High oxidation yields for gas phase reactants and odors
- Complete oxidation to benign CO_2 and H_2O
- Many VOC's and bioaerosols treated
- Low residual ozone
- No chemical additives
- Low energy requirements
- Works in humid conditions
- Long service life
- Negligible pressure drop, if any

But the conventional PCO technology has its own limitations and drawbacks. For residential and commercial applications, it can incur a rather expensive, substantial installation. It is not very effective on all odors, especially since it only treats the air that contacts the activated target as it passes through the unit. We should also note that most UV systems install a glass mercury bulb without protection from breakage. A broken bulb could release mercury which is a potential environmental and health hazard.

Despite its shortcomings, there have been numerous scientific studies done, both pure research and applications, using this conventional PCO technology. Here are some validating comments to reflect on: [10]

> *"One effective method to destroy dilute concentrations of organic and chlorinated organic pollutants in air is heterogeneous photocatalytic oxidation (PCO), which uses a semiconductor catalyst such as TiO_2 and near-UV radiation to decompose contaminates "The large number and variety of chemicals successfully treated by PCO indi-*

cates potentially broad range of application."

John L. Falconer, Ph.D., Professor of Chemical Engineering, University of Colorado

"Photocatalysts for the destruction of indoor air pollutants, including VOC's and gaseous inorganic pollutants such as nitrous oxides, carbon monoxide, and hydrogen cyanide ..." (Heller,1996). "Reports of tests show the technology capable of rapidly destroying toxic components of tobacco smoke such as formaldehyde, acrolein and benzene."

Taken from the American Lung Association website, January 24, 2001

"... The purpose of this study is to investigate the purification of air emissions contaminated with toluene via the heterogeneous photocatalytic oxidation (PCO) process ... Experimental results indicated that near to 100% conversion ratio of toluene are achieved for the initial 30-minute reaction period ..."

Chung-Hsuang Hung, 'Photocatalytic Decomposition of Toluene Under Various Reaction Temperatures.'

" ... Potential applications for using titania-based materials as photocatalysts include ... Destroying volatile organic compounds (trichloroethylene, benzene, formaldehyde, etc). Reducing air pollution in homes and industries such as dry-cleaners, painting booths, and printers ..."

Melanie Louise Sattler, 'Method for Predicting Photocatalytic Oxidation Rates of Organic Compounds'

" ... In addition to automobile exhaust cleaning, use of environmental catalysts such as titanium oxide photocatalysts is rapidly growing for control of residential environments, e.g., antimicrobial activity and odor control ..."

Interdisciplinary Department, Frontier Science Institute, Mitsubishi Research Institute, Inc.

" ... Titanium dioxide is therefore applied for deodorizing, by decomposing substances causing bad odor, and for prevention of air pollution by absorbing and oxidizing ..."

Japan Chemical Week' August 26, 1999

But that's just the conventional photocatalytic oxidation system. As a method of air purification, it is far better than the older UV-ozone or the less desirable corona-discharge (CD) ozone systems since, at least in the view of some, PCO avoids the suspect ozone exposure problem.

...or for much ... to come. The next major development of an ... ion system led to the Photohydroionization® (PHI®) technology. This second-generation revolutionary technology was developed by RGF Environmental Group of West Palm Beach, Florida. They have been the industry leaders and innovators for the past two decades and for that reason, their story is certainly worth telling here.

RGF TAKES THE LEAD

First, we need a brief historical perspective.

Traditional strategies for improving indoor air quality by ventilation and filtration go back a long way. However, developments in the technology of air purification have accelerated during the past two decades. As we saw in the last chapter, the conventional treatment methods of ozonation, ionization and irradiation have all had extensive applications, going back a couple of generations at least. Ozone technology was being used for water treatment soon after the turn of the last century. It later made its way into food and other industries. Ionization and electrostatic precipitators were also in fairly widespread use soon after the Second World War.

Ultraviolet radiation was discovered back in the 1930's and UV germicidal irradiation had been practiced soon after its toxic effects on microbes had been observed. Specifically, experiments with food and water irradiation started earnestly in the 1960's, with promising early results. Direct irradiation of food remains a most controversial issue even today. There is widespread public concern, if not downright fear of its consequences.

In 1985, a new company burst onto the scene with the bold corporate mission 'to provide the world with the safest water, food and air, without the use of chemicals'. That company was RGF Environmental Group, the pioneers of the new advanced oxidation technologies that have now become the space-age leaders in the air purification industry and more.

Their initial interest in the environmental field was in ozone and they made attempts to imitate *indoors* the natural processes by which the sun's UV produces it *outdoors*. Early experiments showed that UV light at 185nm creates a low concentration of ozone. There was some commercial interest in this because it promised a low cost, low maintenance method of producing ozone. The alternative at the time (and even now) is the traditional corona discharge (CD) method which produced a high concentration of ozone. The CD method was impractical because of its high cost, high maintenance and high failure rate. Those were three strikes against it... Out!

But the low concentration of ozone produced by UV radiation had its own nemesis. Twenty years ago, there was no evidence that such low concentrations of ozone would be effective as a purifying or disinfecting agent. Then came the R & D experiments conducted at RGF which demonstrated that the use of UV ozone on industrial waste water was feasible when the low-level ozone was activated with UV light. That process produced hydroxyl radicals, the most powerful friendly-oxidizers that get the job done. **What the ozone lacked in quantity, it made up in reactivity via the powerful hydroxyl radical intermediates.** What does this mean? Just a tiny bit of ozone is enough to produce these aggressive hydroxyl radicals that take over and deliver the purifying results.

Clearly, UV light and ozone were not new discoveries. Even their disinfecting qualities were previously known for sure. Hospitals had used UV light for decades in operating rooms. Ozone was being used to treat water throughout Europe. But what was new about this important find at RGF was the validation of this particular combination for use on water and food as well as air. That shifted the focus to indoor air quality.

This simple breakthrough, for that indeed it was, led to **a string of discoveries** over the next two decades, involving advanced oxidation. These discoveries have resulted in numerous patents and over 500 different RGF commercial products. Since 1985, the company has maintained a steady flow of award-winning, innovative pollution solutions. They employ a Research and Development staff of specialists who are involved in EPA/USDA/FDA/EPRI and University Environmental Studies. Their personnel have had articles published in over 100 periodicals and textbooks.

Today, RGF's air, food and water purification systems are often the industry's choice. They maintain strategic alliances, national accounts and distributorship with many Fortune 500 corporations.

In fact, their impressive corporate client list only reflects the quality of R & D innovation that has been introduced by RGF. Their contribution to the air purification industry, is second to none.

LOW LEVEL OZONE GETS A BREAK

To really appreciate the most recent developments in this field, one must understand that back in the mid 1980s when RGF started their research, air purifiers had started to make their way into the residential market. We know that Ozone systems had been widely used before in the commercial restoration business for fire and flood damage to buildings. Odors and mold were serious problems that the very high concentration ozone treatment seemed to rectify.

These applications utilized the CD system that used a spark between electrically charged plates to simulate lightning.

The problem with these *commercial* CD systems when using air was that with oxygen conversion to ozone, you also get nitrogen conversion to nitric acid and nitric oxide, as we pointed out before. Nitric acid and nitric oxide are so corrosive that maintenance was a real problem. It was costly in dollars, reliability and convenience. Therefore, most professional CD manufacturers also provided oxygen generators (and some still do, by the way) with their systems, to prevent the 'nitric' problem.

But the problems that faced the potential application for *residential* air systems was that the cost of an oxygen generator was prohibitive and the CD units were just used in normal air. That produced high concentration levels of ozone which posed a hazardous threat, in addition to the nitric problem. In fact, ozone readings at the system exhaust exceeded up to 10ppm, whereas federal safety limits were in the order of 0.04ppm. That was a x250 fold excess and could be lethal with continued exposure. While others were content to use CD-air ozone units and hope for adequate ventilation and *voluntary evacuation* while in-use, RGF decided to stay out of the residential market at that time and concentrate only on the ozone commercial market which had restricted use in the hands of professionals and demanded *mandatory evacuation.*

However, RGF also went back to work, back to the development laboratory. Their patience and diligence paid off when, in the late 1980's, they discovered that the lower concentrations of ozone produced with ultraviolet, could have an effect on odors, plus airborne mold and bacteria. Levels as low as 0.02ppm (**which is only half the federal safety maximum**) were demonstrated to have significant results. The real challenge then was to create a residential air purifier that could produce safe, low concentrations of ozone that would not exceed the 0.04ppm federal limits. This was accomplished in the early 1990's, at just about the same time that the U.S. Federal Government was going after the CD ozone residential units with a vengeance. That battle between the Feds and CD manufacturers gave ozone a really bad name, from which the industry is still recovering.

But once again, good science and technology won the day. With the technology to build a device that produced safe, low concentrations of ozone and the ability to ensure that a room would not exceed 0.04 ppm, RGF set out to validate that this device would be effective for mold, VOC's, odors and bacteria. At the same time, *FOX TV News* was doing a three-part series on indoor air problems and asked RGF to test one of the CD units on the market at that time. That was fortuitous. The unit in question produced 18ppm ozone, a potentially lethal amount that drove the camera crew and news reporter right out of the office.

That was the bad news!

The good news came about when the FOX people asked if they could independently test the RGF low-level ozone unit. They ran tests supervised by an independent air specialist and two medical doctors. The results were phenomenal. They could not have been better at that time. FOX ran this on their national *Fox TV News* network and their national *Health News*. What an infomercial that was! So much so, that *Popular Science* picked up the story for their magazine.

Those early units were only the start of things to come. Recall that RGF had gained some experience with low level ozone, using UV lamps. In the 185nm range, these lamps produce low level ozone, just as the sun does. They can be used for air purification under the right conditions. They have low cost, require easy installation, and are effective on mold, smoke, odors and bacteria. The ozone gas travels through the house to provide on-going treatment, unlike the simple UV-germicidal lamps (typically at 254nm) where only the air that passes directly within inches of the bulb gets treated. However, there is a downside because if used improperly, simple UV-ozone units can still exceed federal limits and in any case, ozone is a nuisance and concern to some people. Ozone does not work on all odors and VOC's.

The next major development came about with the exploitation of photocatalytic oxidation. This technology introduced a new weapon to attack air contaminants that proved to be far more effective as an oxidizing agent than even ozone itself.

The RGF Environmental Group utilized the conventional PCO technology and produced units for commercial application. But ongoing research and development in this area led to the much more efficient Photohydroionization® (PHI®) technology that came to market in 2000. The original PHI® cell was designed for use in a central HVAC system but since then, it has been remodeled for application in free-standing units for home or office.

PHOTOHYDROIONIZATION®

Photohydroionization® (PHI®) is the second generation version of the conventional PCO we described earlier. It is an advanced oxidation technology that uses a broad spectrum, high intensity (high energy/efficiency) UV light source, directed onto a multi-metal catalyst/185nm UV target in a low-level ozone and moist atmosphere. This creates an advanced oxidation process that provides an array of friendly oxidizers. These are very safe but aggressive oxidizers that revert back to oxygen and hydrogen.

The PHI® system produces low-level ozone (below the 0.04ppm federal limit), the majority of which is converted into airborne hydro-peroxides and superoxide ions that travel out into the room for complete coverage. The very low level ozone (0.01 - 0.02 ppm), for all intents and purposes, makes the potential ozone hazard a non-issue.

Since all UV bulbs contain mercury, for in-duct installations the bulb inside the PHI®-cell is enclosed in a special polymeric protective cover to prevent any glass breakage or mercury leakage. It has a new, special heavy-duty filament developed by RGF. Combined with a soft start ballast, this PHI®-lamp provides an unprecedented 3-year or 25,000 hour life. (By comparison, current standards for UV bulb-life range from 8,000-10,000 hours).

Not only was a totally new type of bulb used, but a totally new target concept was also introduced with the PHI®-cell. A 360° wrap-around cell of faceted, expanded metal provides increased exposure of the catalytic surface (Fig. 3). The entire assembly is encased in a protective metal cell. It also serves as a germicidal lamp and treats the air that passes through it in that manner.

Fig. 3 A Photohydroionization Cell® (PHI®)

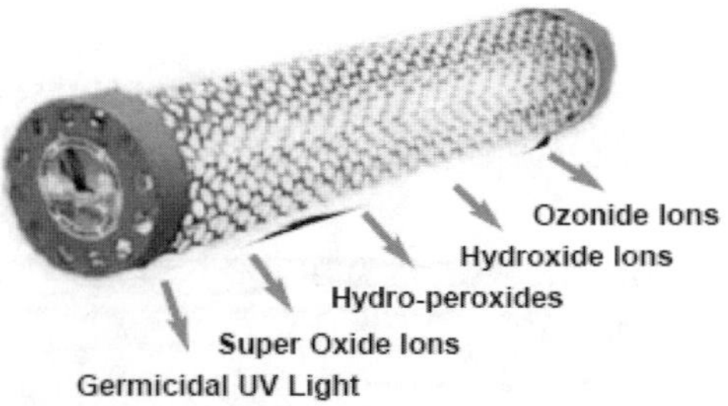

This technology proved to be as safe as it was effective. Mounted in an HVAC system, it provided the whole house with effective air purification. It kept mold from growing on the AC coil and reduced odors, VOC's, bacteria, viruses and mold throughout the house. It required low initial cost, low installation cost, low maintenance and low power consumption. It did all that and still satisfied the federal ozone safety guidelines. It had the broadest range of effectiveness. What more could we ask?

Just wait and see. With PHI® we are still not quite at the summit of advanced oxidation technology. That state-of-the-art privilege is reserved for the process known as **Radiant Catalytic Ionization**™ (RCI™). It is somewhat similar to photohydroionization (PHI®) but differs in a few important details.

Now we can review and describe this innovative third generation technology in more detail.

RADIANT CATALYTIC IONIZATION™

Radiant Catalytic Ionization™ (RCI™) represents the pinnacle of all residential air purification technology known to be available anywhere in the world today. It is based on a proprietary, high-intensity probe or cell, consisting of a broad spectrum UV lamp, surrounded by an open hydrated honeycomb matrix cell that is coated with a superb, proprietary, quad-metallic, hydrophilic coating. Each component has unique features that combine to provide a safe, effective, low-cost and durable unit that should eventually become the ubiquitous solution, making indoor air pollution far less of a problem than it is in the modern industrialized world.

Figure 4. The RCI™ Cell

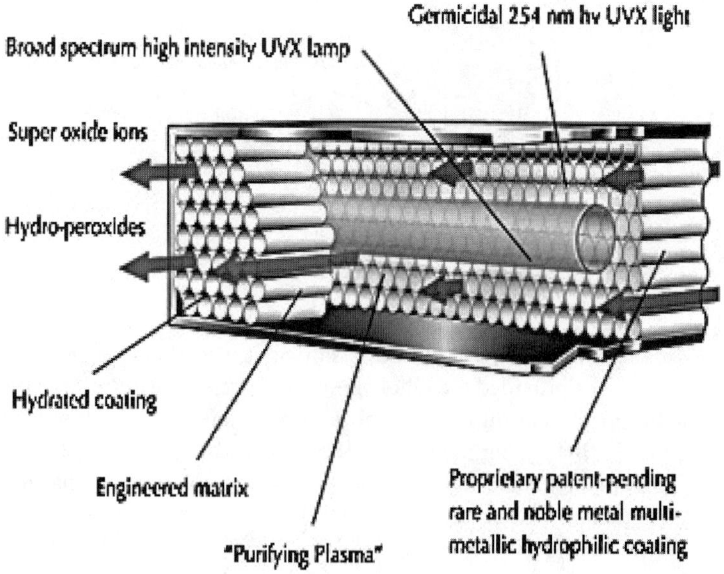

FIVE ESSENTIAL FEATURES OF RCI™

1. **The Honeycomb Matrix**
2. **The Quad-metallic Hydrophilic Coating**
3. **The HE/UV Broad Spectrum Lamp**
4. **The Germicidal Effect**
5. **The Purifying Plasma**

1. The Honeycomb Matrix

This engineered lightweight material and configuration is a characteristic feature of RCI™ technology. It optimizes the area of the catalytic surface and its exposure to the broad spectrum ultraviolet light. That is a major key. The efficiency of the entire RCI™ process is critically affected by the activity at the catalyst surface.

According to Chemical Engineer, James A. Hart, P.E., the RCI™ 'cell has 16.8 times the surface area of the PHI® cell.' In surface catalysis, it is well known that the overall rate of reaction remains constant, but the level of conversion is directly proportional to the catalytic surface area. The catalyst is the same formula for both RCI™ and PHI®. Therefore, it is eminently reasonable to assume that, all other things being equal, the increased surface area of the RCI™-cell will increase the overall yield efficiency by an order of magnitude or more. Therefore, it is only reasonable to expect also that all the earlier validation of PHI®-test results can be extrapolated to the RCI™-cell.

The UV light excites the catalytic surface and in the presence of oxygen, water vapor and airborne contaminates … the magic takes place there … in that local environment. Hence, increased effective surface area is a big plus in realizing the potential of this technology. The honeycomb matrix optimizes the conditions for all the other players to produce at high efficiency.

2. The Quad-metallic Hydrophilic Coating

If the physical geometry of the honeycomb matrix makes for high efficiency of the photocatalyst, it is the coating on its surface that makes it all possible. The RCI™ probe/cell exploits the physico-chemical characteristics of four different rare and noble metals that constitute the very thin catalytic coating on the surface of the honeycomb.

When this proprietary catalyst is excited by the ultraviolet it does a number of things which could be broadly summarized as follows. The quad-metalic coating

- Adsorbs water vapor (is hydrophilic)
- Has particular affinities for organics
- Holds and releases atoms/gas molecules
- Breaks molecular bonds
- Releases electrons
- Steals other electrons.

When all is said and done, the photocatalyst changes everything - but itself remains unchanged. In that regard, it is a true catalyst producing a real 'catalytic' process.

The catalyst is comprised of **four (4) different, rare and noble metals,** each of which makes a unique contribution to the process. The different catalysts do different things. Let's elaborate.

First, is the **titanium dioxide** that we encountered earlier when we described conventional photocatalytic oxidation (PCO). Again, when TiO2 is exposed to UV light of a wavelength below 385 nanometers, in the presence of water vapor, two highly reactive substances are formed: hydroxyl (·OH) radicals and a superoxide ion [$O2^{-1}$]. These are highly reactive chemical species. As we saw earlier, hydroxyl radicals are extremely powerful oxidizers and will attack all kinds of organic materials.

At the **rhodium** catalytic surface, nitric oxide is converted back to nitrogen and oxygen. This avoids the previous nitric problem that we mentioned above. It therefore reduces maintenance problems since it avoids the corrosive action of the oxides of nitrogen.

Silver speeds up reactions at the titanium dioxide surface. It enhances the TiO_2. Research published in the *International Journal of Photoenergy* in 2003 demonstrated the enhanced activity of silver-modified thin film TiO_2 photocatalysts.(11) This silver-doped RCI™ photocatalyst decomposes the typical pollutant x3 times as fast as the undoped TiO_2. It appears that the amount of silver alters the photocatalytic system by enhancing the reduction potential of TiO_2.

There's more. Silver is an excellent electrical conductor but it is also more than that. Studies at the Boreskov Institute of Catalysis published in Russia demonstrated that silver has some unique abilities to work with energized forms of oxygen.(12) It is very stable (hence, its precious mineral value) and will not itself be oxidized in the course of creating and transporting oxidizers. In some ways, elemental silver is a unique catalyst for selective oxidation reactions in the RCI™ process. In fact, we know from some work published in Germany, that four different species of atomic oxygen have been identified, acting in spectroscopically distinct ways with silver.(13)

Finally, there is **copper**. The effect of oxidation state of loaded copper (Cu) on the photocatalytic oxidation reaction has been studied. Reports from Korea demonstrate that copper, just like silver, can improve the functions of titanium.(14) It is expected that the loaded copper may improve the photocatalytic activity of TiO_2.

Again, there is more.

When scientists at the Solar Energy Center at the University of Central Florida looked at solar photocatalytic hydrogen production from water, using a dual bed photosystem, they found that the photocatalysts with the more common copper content evolved the most hydrogen.(15) In a nutshell, the best hydrogen-evolving photocatalyst was copper. It seems clear that copper does have unique abilities to work with hydrogen. This should impart some additional efficiencies to the RCI™ quad-metallic photocatalyst coating.

Altogether, this proprietary quad-metallic catalyst turns out to be **hydrophilic**. That means it has a strong tendency to attract water (H_2O) from the air. This therefore creates an abundant source of hydrogen and oxygen atoms on the coating.

What a choice photocatalyst this turned out to be. It has just the right properties to execute exactly as required. As the people of NASA would say in space parlance, '**it delivers!**'

3. The HE/UV Broad Spectrum Lamp

A critical part of the RCI™ technology that we began describing above is the proprietary bulb that was developed by the RGF Environmental Group. They reworked this component until they obtained a broad spectrum high efficiency HE/UV lamp with the characteristics described in the last section for the PHI®-cell technology. It has a heavy duty filament and emits intense UV light in the 100 - 300nm range. This spans the high energy ultraviolet zones (UV-C plus some UV-V and UV-B).

Very little ozone leaves the RCI™-cell under normal operating conditions.

Specifically, the lamp emits high energy photons at 185nm to create the excitation of electrons into the 'conduction band' of the photocatalyst. From this state, these 'free' electrons can be released to engage in chemical reactions with adjacent molecules, which may or may not be adsorbed on to the surface. This leads to the efficient formation of hydroxyl radicals, superoxide ions, hydro-peroxides and ozonide ions. It also produces low-level ozone in the vicinity of the catalytic surface.

Then comes one of the major tricks of this RCI™ technology. The lamp also produces photons at 254nm which break down the ozone that is formed and returns the oxygen to its normal diatomic state. **The net result is that very little ozone (0.01 - 0.05 ppm) typically leaves the RCI™-cell under normal operating conditions.** Just to compare, normal country or forest air has a 0.01 - 0.02 ppm level of ozone. Most people can smell ozone at about 0.01 ppm or above and the federal safety limit for medical devices is placed at 0.05 ppm. So this very low level of ozone produced by the RCI™ technology makes it eminently as safe as it is efficacious.

But even that is not the whole story. First though, just to finish the description of the RCI™ lamp, we emphasize that in addition to its heavy duty filament, it has a soft start ballast and a protective, insulating poly protective cover (PPC) to shield it from breakage and leakage. Its natural lifetime we pointed out earlier was as high as 25,000 hours.

4. The Germicidal Effect

Inside the RCI™ probe/cell, the broad spectrum UV light rays are also effective at inactivating microorganisms, just as we saw in the last chapter for UVGI. The germicidal UV (254nm) inactivates the microbes passing through the cell by disrupting DNA and RNA. This effectively sterilizes the cell which, for all intents and purposes, can be considered to be 'biologically dead'.

This type of UVGI is very effective when properly applied. As we mentioned earlier, to achieve the proper CT (cytotoxic) value (or "Kill Dose"), you must consider the product of UV energy and time.

UV Energy x Time = Kill Dose (CT Value)

Germicidal UV effectiveness can be easily calculated from some basic principles. The CT values for microbes are well established and can easily be obtained from the scientific literature. The germicidal UV light energy is easily measured with specialized photometers. When designing germicidal UV systems, it is common to use a UV dosage (received by the organism) of 40 millijoules/cm2 (or 40,000 microwatts-seconds/cm^2). This effectively produces a 99.99% reduction in microbes (as measured for example, by sampling the air and counting colony forming units). This is equivalent to a 1 in 10,000 residual population, and is called a '4 log-kill' ie. $1/10^4$. That is an amazing result by any standard!

However, although germicidal UV (UVGI) is a significant component of RCI™ technology, it is not the main technology. It must at least defer that role to the busy photoactivity at the quad-metallic hydrophilic surface and even more.

And still, there is something better yet …

5. The Purifying Plasma

Now then … for the piéce de rêsistance. If all the features of RCI™ technology described to this point were not enough to make it merit the title of 'space-age technology', there is another exceptional feature.

We mentioned above the formation of hydroxyl radicals, super oxides, hydro-peroxides and ozonide ions at the photocatalytic surface, following excitation with high energy (185nm) ultraviolet light. These '***advanced oxidation products***' are very effective oxidizing agents and in part, they react with pollutants in the air passing through the RCI™ probe/cell. That activity alone would go a long way to purifying the surrounding air. But it probably would not go far enough.

We pointed out much earlier in the book that even with the best ventilation and recirculation systems, not all the air will pass through a free standing device or even a central HVAC system.

But here's the most encouraging news ...

The ions that are formed inside the RCI™ probe/cell are released into the air as a 'purifying plasma'. For simple clarification, a 'plasma' is just another word from high school physics to describe a 'highly ionized gas'. This stream of ions escapes into the surrounding air and then travels or drifts by (diffuse) Brownian motion throughout the local environment in search of pollutants. It comprises energized species that can destroy microbes on contact. They are oxidizing agents that can oxidize or neutralize VOC's and other contaminants in the air. As well, they could attach to fine particles to initiate agglomeration and precipitation as the larger 'clumps' fall out the air under gravity. In effect then, **the 'purifying plasma' becomes a collection of efficient 'air scrubbers'**. They go after the air contaminants that remain and wrestle them to the ground.

This is active air purification, independent of the motion of the air actually passing through the RCI™ device itself. This allows the technology to sterilize even surfaces 'across the room.' Whereas other technologies may require bringing the 'pollution to the solution', **with this space-age technology,**

you take the solution to the pollution!

That's the sterilizing power of the purifying plasma!! Therein lies the magic! And yet it's just another example of science in the service of humanity.

Now, you're at the pinnacle - the summit of air purification technology. This is the state-of-the-art as we write today at the beginning of 2006. It may not be the last development in the field, but it is the latest. And just in case you missed any of it, let's summarize the most significant advantages of this space-age innovative technology developed first by the RGF Environmental Group.

TEN ADVANTAGES OF RCI™ TECHNOLOGY

1. **It is the ideal combination of all three conventional purifying methods:**

   Ozonation + Ionization + Irradiation.

2. **The RCI™ Cell/Probe produces and destroys Ozone by the reactions of Advanced Oxidation Products (AOP) at the catalytic surface.**

3. **Typically, more than 87% of the actual ozone produced is decomposed inside the RCI™ cell/probe.**

4. **Devices utilizing RCI™ are therefore not ozone generators in air. The former can actually reduce ozone levels.**

5. **RCI™ develops a local 'Purifying Plasma' of efficient air scrubbers at the surface.**

6. **It uses germicidal UV (254 um) for microbe inactivations.**

7. **It uses the same germicidal UV for Catalytic AOP reactions.**

8. **It uses the broad spectrum combination of UVX wavelengths to produce AOP reactions in air.**

9. **AOP reactions inactivate microbes as well as destroy odors.**

10. **AOP reactants remain effective after leaving the RCI™ cell/probe to function as an *airborne* 'purifying plasma.' They oxidize and sterilize contamination away from the unit.**

Figure 5. The Evolution of Indoor Air Purification

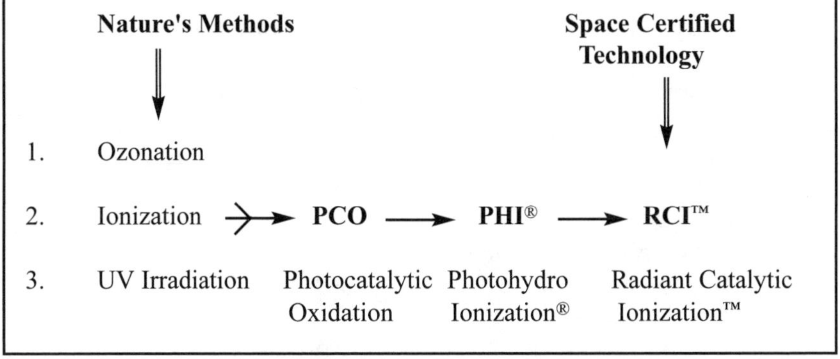

Figure 5 is a summary of the evolution of indoor air purification technology. It began with the conventional application of the three methods utilized by nature for purifying outdoor air and it leads up to the space-age RCI™ technology that we have today - a technology that is effective, economical and safe.

But does it work? Was that your question?

THE RESULTS ARE IN!

In Phase I of a 3-phase program to examine the efficacy of this space-age RCI™ technology in reducing microbial populations, Professor James L. Marsden and his research group at the Food Science Institute of Kansas State University measured the effect on stainless steel surfaces.(16)

They inoculated small, stainless steel coupons (samples) with different microbial cocktails and exposed them in a controlled airflow test chamber at 26° C and 46% relative humidity. The air in the chamber was treated with the RCI™ cell for 0, 2, 6 and 24 hours while the ozone levels were monitored.

All the microbial populations on the stainless steel surfaces tested showed definite germicidal effects, while the consistent ozone level was confirmed at 0.02 ppm during exposure. Longer exposure times resulted in greater reductions with the greatest reductions found after 24 hour exposure. The results for the different microbes are summarized in **Table 12**.

Table 12. Efficacy of RCI at Reducing Microbial Populations on Stainless Steel Surfaces

Prof. J. L. Marsden et al., Kansas State University (2005)

Microbe	**Log Kill (at 24 hrs)**
Staphylococcus aureus	1.85 log CFU/cm2
E. Coli	1.81
Bacilus spp.	2.38
S. Aureus (meth. resistant)	2.98
Streptococcus spp.	1.64
Pseudomonas aeruginosa	2.00
Listeria monocytogenes	2.75
Candida albicans	3.22
Stachybotrys chartarum	3.32

1 log kill = 90% effective
2 log kill = 99%
3 log kill = 99.9%
4 log kill = 99.99%

*** OZONE LEVELS @ 0.02 ppm ***

This preliminary study clearly demonstrates the effectiveness of ozone at these low concentration levels and the effect of the purifying plasma from thc RCI™ cell as a disinfectant tool on surfaces.

Now, look at the data in Table 12 a little more carefully.

The effective log kill on surfaces outside the RCI™ cell ranges from 1.64 to 3.32 after 24 hours. That corresponds to an effective destruction of up to 99.9% of the microorganisms tested. In this field, anything more than 1.0 log kill (i.e. 90% destruction or deactivation) is considered to be an effective treatment.

But there is even more compelling evidence. The same experiments were carried out using a purifier that functions essentially as an ozone generator. The ozone levels in the test chamber were **ten times higher** in that case, yet the effective log kill values for the same microbes were reduced by (up to) 2 orders of magnitude. That means that the RCI™ cell was (up to) **one hundred times** more effective at killing *candida* and *stachybortrys*, to be more specific. The only rationale for that effect on **surfaces outside the cell** is that the **purifying plasma did its job with distinction!**

It took the Solution to the Pollution!

One further point is noteworthy. These experiments were extended only to 24 hours of exposure. In normal real-life applications, the RCI™ cell can and should be utilized 24/7. In other words, with extended exposure, the log kill could, in principle, go very much higher. That implies that sterilization could be extrapolated to almost 100% completion.

This may have potential applications in food processing, especially after the U.S. FDA approved in June 2001, the use of ozone as a sanitizer for food contact surfaces, as well as direct application on food products. But it also does confirm the value of this technology in normal room air purification where surfaces in the room are subject to incidental bio-contamination.

This research is on-going. In Phase II of the project an aerosol study is being conducted to see the efficacy of the RCI™ technology in reducing microbes suspended in the air.

PUTTING IT ALL TOGETHER

In summary, there are serious *concerns* about the quality of indoor air today that we simply can't ignore. We have a **Problem** that the U.S. EPA considers one of the most significant environmental issues today but one that remains, by and large, below the popular radar screen. We spend 90% of our time in buildings designed to lock us in with recycled *contaminants* that we can't avoid. But the health and other *consequences* are things we can't afford.

Therefore, in pursuit of a ***Solution*** to provide **Fresh Air for Life**, we have outlined an approach to a solution of the indoor air problem based on **FIVE ESSENTIAL STRATEGIES:**

1. **Eliminate** all known sources of indoor air contaminants.

2. **Ventilate** the air space to dilute and replace polluted indoor air.

3. **Filter** as much of the particulates, bioaerosols and volatiles as possible.

4. **Purify** the air by classic low-level ozonation, ionization and UV radiation methods.

5. **Use the space-age Radiant Catalytic Ionization™ technology** to exploit the synergy of all the purification methods and more.

In Part Three of this book, we will discuss the ***Application*** of these **FIVE ESSENTIAL STRATEGIES** in practical situations. We will go from protecting your personal space and maintaining a healthy home, to avoiding the pollution associated with transportation, the workplace and other commercial environments.

This should allow you to experience **Fresh Air for Life** and help you win in *your unseen battle with Indoor Air Pollution.*

SPACE CERTIFIED TECHNOLOGY

We conclude this chapter just where we began it: in space. The Radiant Catalytic Ionization™ technology which had its origins in photocatalytic oxidation has earned the prestigious designation as a Space Certified Technology awarded by the U.S. Space Foundation.

Founded in 1983 and headquartered in Colorado Springs, Colorado, the Space Foundation is a national non-profit organization that vigorously advances civil, commercial and national security space endeavors and educational experience. Its Board of Directors represent a partial list from among the Who's Who in the field of space technology, administration, corporate leadership and media promotion. It includes astronauts, scientists, politicians, business executives and policy makers. The Space Foundation has offices in Washington, DC and Cape Canaveral, Florida. It annually conducts, along with its partnering organizations, the *National Space Symposium* in the Spring and *Strategic Space* in the Fall.

In cooperation with NASA, the Space Foundation developed the Space Certification Program as a unique branding program to promote the extraordinary products and services that bring the benefits of space-related technology home to earth and improve the quality of our lives.

The Space Certification Program is designed to certify products that are a direct result of space-related technology, or that use space-derived resources for consumer benefit. This exciting program highlights products in three distinct areas: technology, education and entertainment. Of these, RCI™ clearly falls into the technology area.

Radiant Catalytic Ionization™ (RCI™) is the only air purification technology for residential application that has the distinction of bearing the Space Certified seal of the Space Foundation.

You need not wonder why! At least, not anymore.

Now, we turn in Part Three of the book to go beyond the **Problem** (underscored by the fact that *Untold Thousands are dying annually from Indoor*

Air Pollution) and the **Solution** (underlined by the knowledge to '*Use these* ***FIVE ESSENTIAL STRATEGIES*** *to Purify your Indoor Air*') to the **Application** (undertaken as a challenge of '*How to Purify* ***YOUR FIVE IMPORTANT SPHERES*** *of Activity*').

There are more exciting discoveries ahead.

"All of us face

a variety of risks to our health

as we go about our day-to-day lives....

Indoor air pollution is one risk

that you can do something about."

— U.S. Environmental Protection Agency

Part III
The Application

Purify your **FIVE IMPORTANT SPHERES** of Activity

ELEVEN

Defend Your PERSONAL Space

"If you can't breathe purified air, nothing else matters"

Some things in life appear too good to be true. And many of those apparent wonders are just that ... illusory at best. They seem so good from afar, but on closer inspection, they really are far from good.

"It's hard to resist those high-tech gizmos you see in some catalogs. You know, those strange-looking gadgets that promise to ease your aches and pains, help you look younger or even make your clothes fit better. Sometimes they work, but some of these 'miracle cures' can turn out to be snake oil."

That's how Hoda Kolbe introduced her report with a *Dateline NBC/Good Housekeeping* exclusive when they put a few of these 'gizmos' to the test, including a 'beeper-like gadget to clean your personal breathing space'. They could hardly believe that a personal air purifier that works on a 'revolutionary technology' could destroy pollutants in the air and cleanup your own breathing zone.

One *Good Housekeeping* staffer had awful cat allergies that would make her downright miserable. She would have itchy eyes, an extreme itch in her nose and would keep sneezing uncontrollably whenever she was exposed to cats. So she was given one of these incredible gizmos to wear around her neck and amazingly - even after more than half an hour in a cat adoption center, surrounded by 18 cats at the same time - she did not sneeze even once. Was it really the purifier she was wearing that made the difference? Perhaps it was just mind over matter, a mere placebo effect.

That was not good enough. *Good Housekeeping* turned to their engineers for help and they devised a "smoke test" to see if this air purifier could indeed clear out a tank of smoke. Here's what they did. They first lit a cigarette and allowed it to burn inside a tank to build up a lot of smoke. Then they put the turned-on air purifier inside the same tank and watched as the streams of smoke declined until nothing remained visible. The effect was demonstrable for all to see.

Perhaps in this case, it was indeed very good, but not 'too good' for it was also true. That personal air purifier featured on *Dateline*, the *CBS Evening News, NBC News, Oprah,* the *Good Housekeeping Institute* and even in *The Wall Street Journal* could really deliver. It has been proven that a personal air purifier can be exactly that - thanks to an ingenious engineer, Stanley Weinberg, the inventor and founder. It does in fact eliminate fine particles, mold, allergens, bacteria and viruses from one's breathing zone.

Needless to say, the **First Sphere** of application of any technology to address the problem of air pollution should begin exactly at this level. You should be concerned about your unique and personal environment which surrounds you 24/7. As was pointed out in Chapter 1, air is the natural indispensable human habitat. Constant breathing is your most intimate interaction of any kind. What you breathe in is derived from your immediate surroundings - not in the atmosphere outside, and not even across the room. It is in that close personal space, the air moving across your face at any given moment, that dictates what is inhaled into your lungs, destined to enter your bloodstream and travel throughout your entire body. That adjacent air, in that private space, becomes your personal source of life and health.

You must protect your personal space, even if you may not be able to do much about the state of atmospheric pollution outside. Of course, every little bit counts and personally responsible lifestyle choices can make a collective difference. But the wheels of industry and the emissions of thousands of local transportation vehicles will probably override whatever reductions you can personally achieve. So even after you've done *your best* for the environment, you will still have to contend with air pollution. There's no escaping it.

But as you also learned earlier in this book, the problem of air pollution is primarily an indoor problem. The U.S. EPA, the American Lung Association and all responsible agencies and institutions that are responsible and aware, come to the same conclusion. Indoor air is typically several times more polluted than the atmosphere outdoors. Also, recall that we usually spend about 90% of our time indoors, trapped in our energy efficient buildings which have sealed in the air contaminants 'with deadly precision.'

However, after you have *eliminated* as many known sources of indoor contaminants as you can; after you have increased *ventilation* to dilute and even replace the polluted air inside; after you have employed *filters* to remove what can be blocked mechanically or by adsorption, there will still remain contaminants in the indoor air that you cannot avoid by all those earlier strategies. You will be obliged to resort to air purification methods that deal with the residual *contaminants* that still pose significant challenges and threaten health *consequences* that you can ill afford.

You will need to protect your space - your personal space. That is the First Sphere of activity to address, by way of application of any air purification solution. You need to, and the good news is that you can!

Thanks again to the innovative ideas and creative genius of Stanley Weinberg, a personal air purifier has been developed that utilizes some of the same technologies outlined in Part Two of this book. This type of device has been developed, patented and researched so that today, some of the best peer-reviewed scientific data in the industry has been published on the effectiveness of this device in destroying a wide variety of airborne pollutants. These include airborne pathogens (disease-causing microbes), chemical toxins and allergens, and fine and ultrafine particles that occupy the immediate breathing zone of many individuals.

A PERSONAL AIR PURIFIER

This technology stands on a solid scientific foundation. The inventions are adequately covered by the issue of at least **two U.S. Patents** and the applications are rigorously demonstrated in the publication of no less than **13 peer-reviewed journal articles** to date.

Apparently, the invention was motivated by an effort to alleviate Mr. Weinberg's personal battle with allergy and asthma attacks. In the process, he not only solved his own problems in that regard, but made available in our generation a real possibility for improved self-defense against what has become an increasingly serious threat. This can even have new significance, with the imminent dangers posed by common infectious diseases, probable pandemics and even senseless bioterrorism.

The best way to describe the personal air purifier is probably to just quote the Abstracts of the U.S. Patents as issued, which are now available in the public domain. Here's the first one:

Inventor: Weinberg, S. **US Patent No. 5,484,472**
January 16, 1996

Miniature Air Purifier

Abstract

A small, battery-powered air purifier can be clipped to a wearer's front shirt pocket or worn suspended from a cord about the wearer's neck. The device includes a housing containing a compact circuit that transforms direct current provided by the battery into a negative high voltage pulsating current which is connected to a sharp metal point contained within a chamber inside the hollow housing. A corona discharge forms on the sharp point, ionizing air molecules and any particulates, and generating ozone. An opening into the chamber is covered by a non-corrosive metal grid connected to the positive terminal of the battery. The negative ions are attracted to this grid, thereby completing an electrical circuit. Movement of the ions to the grid results in mass movement of air which causes a stream of air to emerge through the grid. As the air passes the grid, negatively charged particulates are deposited on the grid. The cleansed air stream, containing traces of ozone and negative ions, can be directed to flow across the face of the user, thereby limiting the contact of contaminated ambient air with the eyes and nasal passages of the user. An activated charcoal filter pad can be attached to the device to interact with the cleansed air stream to reduce the ozone level.

This earlier miniature air purifier was characterized by its compact size, allowing it to be worn clipped to one's clothing or on a cord about one's neck. However, the need to incorporate a fairly complex electronic circuit board, including two inductor components, resulted in the unit having a dimension of at least 2.5" x 4.0" x 1" and weighing at least 7 ounces. The circuit of that device required a significant current flow from the battery, which considerably limited the battery life. Typically, the unit would operate for about 10-15 hours using a fresh alkaline battery.

You must protect your personal space ... and the good news is that you can!

The primary routes of air purification in this early purifier were chemical destruction of pollutants by ozone (ozonation) and the corona discharge, and the electrostatic precipitation of particulates as a result of charging by negative ions (ionization). The primary source of airflow was a mass air flow due to acceleration of negatively ionized air molecules towards a grounded metallic grid.

Because ozone can be irritating to the respiratory system, it was desirable to provide a more effective corona discharge for destroying a wider range of chemical pollutants while, at the same time, limiting the amount of ozone emitted. There was still considerable need for a miniature air purifier that could be very efficient at producing a corona discharge for better removal of pollutants and longer battery life, with minimal emission of ozone.

Therefore, the inventor went a step further in refining and redesigning the device to meet a number of different objectives:

- To provide a miniature air purifier that could operate for at least 20 hours on a typical 9V-alkaline battery
- To simplify the electrical circuit using only a single inductive component so as to reduce the size and weight, as well as the cost of the device, and to reduce the generation of any radio frequency interference
- To optimize corona discharge purification to maximize killing of pathogens, detoxification of chemical pollutants, and neutralization of allergens, while minimizing release of ozone
- To provide a small wearable air purifier that includes an alternative embodiment that also pre-cleans the air by means of pollutant and pathogen filters prior to the corona discharge

All these objectives and more were met and the new portable personal corona discharge device was also patented. Here again is the self-explanatory patent abstract:

Inventor: Weinberg, S. **US Patent No. 5,667,564**
September 16, 1997

Portable Personal Corona Discharge Device
for Destruction of Airborne Microbes and Chemical Toxins
Abstract

A miniature air purifier produces a corona discharge surrounding a needle-like emitter point connected to a novel negative 8,000-volt DC power supply. The power supply operates from a nine volt battery and contains a step-up voltage inverter having a single transformer outputting high voltage spikes with a voltage multiplier operating on the output of the inverter. The production of high voltage spikes of about 200 Hz rather than a sinusoidally varying voltage significantly reduces current consumption. The needle-like emitter point is located about 1/4-inch from an 80% open mesh metallic grid held at ground potential. Corona discharge at the emitter point ionizes the air and creates ozone, and nitric oxide both of which combine with direct electron impact decomposition to detoxify and destroy a wide variety of airborne pollutants including pathogens, chemicals and allergens. The grid attracts negatively ionized air molecules thereby creating a flow of purified air out of the device and also provides a surface for electroprecipitation of ionized particulates. An alternative embodiment uses a miniature brushless DC fan to draw room air through a pathogen and pollutant removing filter. The filtered air is exposed to the corona discharge for additional purification and then accelerated out of the device by the emitter point and grid.

What's the bottom line? Ignoring the physics, this latest personal air purifier is compact, about the size of a small beeper and weighs only 4 ounces. It is easily portable, worn around the neck or clipped on to one's clothing. It operates on lower voltages (8kV) which still yield a vigorous corona discharge that makes 30,000,000 electrons per second available for ionizing and destroying pollutants through electron impact.

Note the intensity of this ion stream (ion emission rate). Needless to say, the effectiveness of this device must be conditional on this quantitative output. Any imitation of this technology that trivializes this significant ion stream might be an impressive-looking toy, but **not** an effective air purifier. Perhaps the first question to be asked of any such device that would be a 'personal air purifier' would be "How many ions does this device put out?" It must be of this 10^7/sec order of magnitude.

The mass air flow rate streaming from the device is about 50 feet per minute which is just about right to reach the user's face. At lower flow rates, the expelled air may not get there and at higher flow rates, there is insufficient residence time in the CD to obtain optimal destruction of pollutants.

Most importantly, the level of ozone produced is less than 0.04 ppm and therefore within all acceptable safety standards today. In experimenting with the prototypes of this invention in an effort to balance air flow and battery life, there was a surprising discovery concerning ozone generation. When the current flowing into the high voltage generation system is limited and the high voltage itself is kept preferably at no more than about - 8kV, the level of ozone drops dramatically without significantly impacting the mass air flow at the grid. This reduction of ozone production results in the electron impact (ionization) effect of the CD predominating.

The ozone produced by the discharge is chemically very active. It oxidizes many pollutant chemicals and also inactivates pathogens and allergens. However, most of the ozone produced inside the device is actually consumed in the oxidative chemical reactions that take place *inside* the device as the pollutants are neutralized.

Besides ozone-mediated oxidation, there is tremendous destruction of airborne chemicals by direct electron (plasma) impact decomposition within the CD. Particulates passing through the discharge without being incinerated become charged, so they precipitate out on grounded surfaces like the metallic grid. Alternatively, they pass through the grid to be deposited on any surface in the room.

A specially selected hot melt adhesive is used to conformationally coat the high voltage circuits to prevent inappropriate corona discharge and to encapsulate the system against moisture and other environmental factors. This treatment significantly reduces energy requirements, further adding to overall increase of battery life. This encapsulation results in much lower finished device weight (an important factor for wearability) compared to conventional epoxy potting devices.

THE PROOF OF THE PUDDING

A device such as this does really sound too good to be true. But the hallmark of all good science and technology is the demonstrable reproducibility. Different investigators at different times, in different locations can do the same test or experiment under similar conditions and get the same results. **That's science** (plain and simple - and not *scientism* or *scientology*). The personal air purifier has been subjected to thorough testing for both safety and effectiveness in recognized laboratories and the results speak for themselves.

Safety

The principal concern over the use of an ionic air purification device that produces ozone from a corona discharge (in addition to the unipolar ions) would clearly be the level of ozone produced, since above recognized standard levels, ozone can be a respiratory irritant. Therefore, **ozone levels produced by the personal air purifier** developed by Wein Products, Inc., **have been tested and proven to be safe** by Underwriters Laboratories Inc. (USA), Fuji-TV Product Research Testing Laboratory (Japan), the University of Southern California, and Advanced Pollution Instruments (USA), among others.

For example, at the independent Underwriters Laboratories, the maximum ozone concentration measured during the first 24 hours of operation of the personal air purifier was 0.026 ppm, in compliance with Section 37 (ozone test) of UL867. Similarly, tests conducted at Fuji-TV Product Research Laboratory (Tokyo, Japan), an independent and reputable test lab agency in Japan, showed that the same device produced only 0.033 ppm of ozone.

A third quotation may be added from another independent organization testing for safety and performance, which is Intertek ETL Semko (Europe/North America). They tested the personal air purifiers in accordance with Section 32.1 of the standard for Safety of Household and Similar Electrical Appliances Part 2-65: Particular Requirements for Air Cleaning Appliances (IEC 60335-2-65). They found the unit generated a maximum of 0.032 ppm of ozone after 24 hours of continuous operation in the center of a room approximately 8'x12'x10'. The dimensions were not significant enough to impact the test results.

In all cases, the personal air purifier was demonstrated to produce less ozone than the maximum (0.05 ppm) safety standard allowed by any federal agency.

Effectiveness

Eminent researchers in air pollution science have studied the effectiveness of the personal air purifier and published their findings in **more than a dozen articles in peer-reviewed journals.**(1-12) Work done at the Department of Environmental Health, Center for Health-Related Aerosol Studies, **University of Cincinnati**; at the Department of Microbiology and Molecular Genetics, **UCLA**, California, and at the Departments of Physics and Astronomy, and of Chemical and Biochemical Engineering, **UC Irvine**, California - these three centers, primarily - has unequivocally established that the device in question can substantially prevent the inhalation of toxic particles such as smoke, dust, pollens, molds, many allergens, fungi, germs and the most dangerous particles that would otherwise remain trapped in the lungs.

For the reader who may still be thinking that a personal air purifier is still 'too good to be true,' let's just present some highlights of this irrefutable evidence.

The Effect on Airborne Bacteria

To test the effect of the personal air purifier on bacterial survival in order to establish a scientific basis for its potential beneficial application, Professor Imke Schoeder (Department of Microbiology and Molecular Genetics, UCLA) and Dr. Alan Spira (Medical Direcor of the Travel Medicine Center) studied the non-pathogenic bacteria *E. coli* and *Paracoccus denitrificans,* which are model organisms for airborne pathogens.(13)

First, a testing system was devised whereby bacteria were blown towards growth media-filled Petri dishes and shown to consistently grow on the plates when aerosolized by their system. That system was meant to serve as a model of humans inhaling bacteria, whereby the plates served essentially in place of the nose and mouth. The purifier was then placed under the bacteria-laden air stream, and a greater than 95% reduction in bacterial colony counts was observed, on the plates. Controls without the active CD were used to demonstrate that the bacteria-laden air was not simply blown away as the reason for lowered colony counts. Another control included using Petri plates spread over the whole chamber, to see if scatter or deflection of the air stream might be another reason for colony count reduction, but this was also shown not to be the case.

These experiments substantiate the conclusion that the personal air purifier does indeed reduce bacterial growth.

Classically, bacteria are categorized by shape and certain metabolic characteristics. Among the most common categories by shape are *rods* and *cocci* (grape-like). In the aerosolization experiments just described, *E. coli* is representative of rods and *P. denitrificans* represents the cocci. They are both very safe to handle in the laboratory. These results can probably be extrapolated to suggest that other rods, like pathogenic *E. coli* which is so destructive to the human gastrointestinal system (called enterohemorrhagic *E. coli*, EHEC), and other varieties of microorganisms such as the airborne pathogen *M. tuberculosis*, will be similarly aerosolized and effectively removed from the breathing zone when the personal air purifier is used.

The personal air purifier does indeed reduce bacterial growth.

In fact, the test of the effect on pathogenic *E. coli* 0157:H7 was done at Omni Tech Laboratories, Inc. (Marietta, Georgia) and a 3 log kill (99.9% reduction) was observed on polyethylene board surfaces within 10 minutes of exposure from a personal air purifier. Even more impressive was a 5 log kill of that same organism on an agar plate, within 5 minutes exposure. Similar results were obtained by Omni Tech on *Salmonella* and *Listeria* species bacteria.

Other impressive results were obtained by Olga Gornostaeva and coworkers at the University of California, Irvine. They observed a typical 4 log kill (99.99% reduction) of E. coli strain SC29Q2 on the surface of agar plates, after 10 minute exposures at 3" distance from a similar sanitizer apparatus from Wein.(14)

Removing fine and ultra fine particles

In an attempt to understand the air cleaning mechanisms for the personal air purifier, Dr. Grinshpun and his colleagues at the Center for Health Related Aerosol studies, University of Cincinnati, investigated the effect of the device on the aerosol concentration measured in the vicinity of a human manikin placed in a relatively small 2.5 cubic meter walk-in environmental chamber.

Most respiratory problems are closely associated with fine (less than or about 2 microns) and ultra fine (less than or about 0.1 micron) particle size fractions. These are the particles that penetrate deep into the lung tissue and in some cases, even end up in the blood circulation.

In their pilot study, the researchers used three types of respirable particles: polydisperse NaCl (salt) particles ranging from 0.3 - 3.0 microns, monodisperse PSL spheres and *Pseudomonas fluorescens* bacteria of typical 0.8 micron size.(5) These test particles represent the size range of microbial fragments, single bacteria, most fungal spores and microbial aggregates. With the air purifiers generally positioned on the chest of the manikin, the experiments were conducted under calm air conditions with breathing and non-breathing manikins, as well as in mixed air.

The results clearly demonstrated up to 100% particle removal efficiency in less than 2 hours, which was not affected by either the size range of the particles or the particle type (biological or non-biological). Air mixing increased the efficiency but the manikin's breathing had little or no effect. The unipolar ionic air purifier seemed to be quite efficient in reducing aerosol exposure in the breathing zone when used in confined spaces characterized by a high surface to volume ratio.

In a follow-up study, the Cincinnati group also looked at ultra fine particles because several bioaerosol agents that cause emerging diseases, as well as those that can be used for biological warfare or in the event of bioterrorism, belong to this particle size range. For example, SARS is caused by a *coronavirus* (approximately 0.1 micron) and anthrax by *bacillus anthracis* (approximately 1.0 micron). They measured the aerosol concentration, aerodynamic particle size distribution and the particle electric charge distribution in real time in a larger volume chamber, about the size of a typical room (20-40 cubic meters).[1,8]

The results of this latter study again demonstrated that **the personal air purifier is very efficient in controlling fine and ultrafine aerosol pollutants in indoor air environments**. The efficiency depends on the ion emission rate, since this affects the particle mobility. (Recall what was said earlier about this acid test of any personal air purifier like this. The rate must be in the 107/sec range.) The surface-to-volume ratio is also important. With increasing air volume, more time is needed to reach a certain air cleaning level. Thus, this type of personal air purifier would be especially efficient in more confined spaces.[3]

It is clear that with the unipolar ion emission device, the aerosols or particles in the air environment become charged with like polarity and therefore repel each other out of the breathing zone when it is worn by the individual. The combination of the bactericidal effect above and the decrease in the aerosol concentration can together significantly reduce human exposure to indoor air pollutants, such as particles and microorganisms.

Experts who peer-reviewed these studies made the interesting observation that the physical concentration reduction of these ultrafine particles does not depend on their infectivity as long as it is affected by the forces of electrostatics, diffusion, turbulence and gravity. In fact, **even if the enemy riding on these tiny aerosols was the most lethal pathogen known to man, it would still be controlled by these same physical forces, and it would still be reduced in the breathing zone as the aerosol concentration is obliged to.** The defense here is at least to remove the pathogen from the breathing zone and not even to necessarily destroy it.

Think about that.

ENHANCING CONVENTIONAL PROTECTION

The conventional method of human protection against airborne contaminants has for a long time been the filtering-face-piece mask. Millions of workers worldwide - including healthcare personnel, industrial workers exposed to dust, volatile chemicals etc., relief and sanitation workers who must contend with odors and fumes, and many other dedicated public servants - all routinely wear such respirators in their workplace. Responsible cleaners in the home using toxic aerosol sprays, disinfectants and the like, as well as remodellers and renovators, make wearing this disposable protection device a standard procedure on the job.

At least two types of respiratory masks are commercially available from major manufacturers. One is the NIOSH (U.S. National Institute for Occupational Safety and Health) certified N95 respirator and the other is the conventional disposable surgical mask. The N95 respirator consists of inner and outer cover webs made of synthetic rayon. Its filter is made of polyester and polypropylene, with the electrostatically-charged microfibers providing relatively high filtering efficiency. In the case of the surgical mask, the polypropylene filter is sandwiched between inner and outer webs made of rayon. The filter of the surgical mask actually has lower filtration efficiency compared to that of an N95 respirator.

The filtration efficiency is dependent not only on the filter properties but also on the particle size. For example, a Type N95 respirator may allow up to 5% penetration in a 'worst case scenario', when most-penetrating NaCl particles of 0.3 micron mass median aerodynamic diameter are drawn through the filter at a rate of 85 lit/min. However, larger bacteria like *M. tuberculosis* may have a penetration efficiency through the same mask at strenuous workload of only 0.5%. That's an order of magnitude difference. The face-sealed filter of a conventional healthcare mask, will provide relatively low pressure drop and consequently a good comfort level. This would allow approximately 15% of airborne *Mfb* surrogate bacteria to penetrate, thus providing only 85% protection against these bacteria. In addition, the use of an improperly fit-tested, tight-fitting mask (which is not uncommon) may allow additional particle penetration through the face-seal leaks, further decreasing the level of respiratory protection.

With all this in mind, there is clearly room for improvement in the filtration efficiency of existing masks, if the comfort level provided by these simple barrier devices could remain the same. That's where the ionic emission personal air purifier can make a huge difference.

The same group at the University of Cincinnati has now further developed and built their sophisticated laboratory for evaluating various respiratory devices with bioaerosol particles and their surrogates, using a manikin-based protocol in a 25 cubic meter indoor chamber. The aerosol concentrations are measured in real-time inside and outside the mask worn by a manikin. The measurements are conducted by a particle selective aerosol spectrometer. The penetration efficiency is determined as a ratio of these concentrations for specific particle size fractions at different breathing flow rates.

They have demonstrated the value of this novel concept by evaluating the effect of the air purifier on the penetrating concentration and size distribution of fine and ultra fine aerosol particles, through a respiratory mask face-sealed on a breathing manikin in their 24m^3 indoor test chamber. They exposed the manikin to polydisperse surrogate aerosols that simulate viral and bacterial particles into the submicron range, with respect to their aerodynamic size.(2,4,11)

The results clearly showed that the particle penetration through either mask tested was decreased by one-to-two orders of magnitude as a result of the unipolar ion emission in the chamber. The flux of air ions migrated to the breathing zone and imparted electrical charges of the same polarity to the aerosol particles and to the respirator filter surface. This created **an electrostatic shield** along the external surface of the filter, thus enhancing the protection characteristics provided by the respirator.

To illustrate this effect in numerical terms, the researchers considered the following example: (2) An individual exposed to the influenza virus concentration of 10,000 per cubic meter, inhales approximately 4500 x 0.18 = 810 viruses during a 15 min period when breathing through a surgical mask at 30 liters per minute in the absence of ion emission. The respirator penetration efficiency (Ep) for 0.1 micron particles with the surgical mask was measured at 0.18 or 18%. The continuous emission of negative air ions (or 10,000 electrons per cubic centimeter) in a 25 cubic meter room would reduce the indoor viral concentration by a factor of 9 during that 15 min interval. In addition, it would enhance the surgical mask protection, reducing Ep from 18% to at least 0.19%. Thus, only about (4500/9)x0.0019 = 0.95 or 1 virus would be inhaled in 15 minutes. Given that the infectious dose of *Influenza A2* is 790 viruses, **the ion emission effect would make a huge difference with respect to the health risk**. Amazing!

In a follow up study, the same group at the University of Cincinnati looked at four types of half-mask face piece filtering devices, operating at two different breathing flow rates, and tested them with unipolar ion emitters again, exhibiting different emission rates and polarities.(4) For the targeted particle size range of 0.04 - 1.3 microns, a 12-minute air ionization in the vicinity of the manikin enhanced the respiratory mask performance by a factor ranging from 1.61 to 3,250, depending on the respirator type, breathing flow rate, and the ion emission rate. The effect was achieved primarily within the first 3 minutes.

That's even more powerful stuff!

This performance enhancement effect can, in principle, prove to be crucial in minimizing the infectious risk in all cases when the conventional filtering-face piece respirators are not able to provide adequate protection against indoor viruses and bacteria.

Many healthcare workers who are unsuspectingly exposed to numerous airborne infectious agents on a daily basis today, may well benefit from this additional enhancement. It would make consummate sense, it seems to me, for the typical doctor, (surgeon even), nurse, paramedic, personal support worker, clinic and hospital staff, immigration personnel, and all others who routinely are at risk of encountering unforeseen airborne infection, to utilize this added protection routinely **NOW**. But (and God forbid!) in the event of a pandemic breakout of some threatening infectious agent (and epidemiologists are already warning of the imminent Avian bird flu possibility), or in the other nightmare scenario, with a successful bioterrorist attack in a major urban area (that is no longer a Hollywood fantasy, especially after 9/11) - there could be an immediate need for millions of readily available respiratory protection devices which could each be enhanced, again in principle, by an ion emission personal air purifier.

That has big league implications!

Perhaps SARS was a tongue-in-cheek welcome wake-up call. The sight on television of frantic ordinary people in major cities around the world wearing disposable respiratory masks in the streets and on subway trains has hopefully awakened both the general public and policy makers regarding the serious implications of the infectious threat on a massive scale and what adequate protection would demand.

You need not wait. You can now take your personal and independent proactive step in whatever degree of self defense is now available to you. That might well include a functioning personal ion emission purifier such as we have just described. You could have such a private companion to daily protect your breathing zone.

Everybody needs a 'buddy' just like that! Perhaps you'll get even more than you bargain for. You might even be surprised to learn, that you would also enhance your protection against the most prevalent chronic disease in America. Guess what?

THE #1 CHRONIC DISEASE IN AMERICA

In September 1999 the Mayo Clinic published a watershed article in their *Proceedings* journal in which they claimed the discovery of the cause of the most common chronic disease in America.(15) No, it was not a cause for diabetes, heart disease, arthritis, hypertension or cancer. It was the cause of chronic sinusitis.

It is estimated that 37 million Americans suffer from this annoying condition. Chronic sinusitis is an inflammatory disease of the nose and sinus cavity that lasts three months or more. It is contrasted with acute sinusitis which typically lasts a month or less and is usually associated with a bacterial infection.

People who suffer from chronic sinusitis have common symptoms of runny nose, nasal congestion, often loss of smell and headaches. Frequently, the chronic inflammation leads to polyps, those small growths in the nasal passages which hinder breathing.

Chronic Sinusitis - 37 million Americans suffer from this annoying condition.

Up until this Mayo paper, the cause of chronic sinusitis was often associated with allergies and perhaps chronic bacterial infection. Fungus allergy was presumed to be involved in less than ten percent of cases. However, the Mayo group discovered fungus in 96% of over 200 patients' mucus. They identified a total of 40 different kinds of fungi in these patients. Also, blood tests done on about half the patients who had had surgery to remove nasal polyps, revealed evidence of eosinophils (a type of white blood cell activated by the body's immune system) in about 96% of these patients.

The explanation that emerges is that of **an immune system response to fungi**. In sensitive individuals, the body's immune system produces eosinophils to attack fungi and these irritate the membranes in the nose. The mechanism of formation of chronic sinusitis would therefore follow a pattern something like this: You first inhale particles (mainly fungus or mold) in the air, to which you have an allergic reaction. This early reaction causes small pits to form in the membranes that line the sinus pockets. These pits become traps for mucus so that the normal mucus drainage is impeded. The stagnant mucus then gets infected, which causes nasal polyps and thickening of the nasal sinus lining, which further obstructs the mucus outflow. The polyps cause more infection and the infection causes more polyps and there you have it, a vicious and self-perpetuating cycle ensues.

But the key is the initial hypersensitivity (probably type 2) reaction to the mold antigen. All type 2 hypersensitivity reactions stop when the antigen is removed. It's therefore not surprising now to observe that twenty years of experience have demonstrated that when patients thoroughly cleaned their environmental air, their sinuses improved.

Otolaryngology (ENT) specialist Dr. Donald Dennis published some interesting results on the treatment of allergic fungal sinusitis by decreasing the environmental fungal load with the same personal air purifier we have been discussing.(12)

The air mold level conducive to good health was first determined by testing a one-hour gravity plate exposure inside each chronic sinusitis patient's home, followed by endoscopic photographs as remediation of the mold was done. Over 300 patients' homes were covered and it was found that a mold count of 0-4 colonies with a one-hour gravity plate exposure was required for the sinus mucosa to clear as evidenced by endoscopic photography.

The personal air purifier was tested using a manikin head with a hose (through the nose) attached to a suction pump and placed in a jar with a mold plate at the bottom. The results demonstrated that the ion emission device reduced the mold load from 'too numerous to count' (TNTC) to a single colony in 2 hours. This load level is 800% better than the standard required to achieve sinus health.

Up until most recently, antibiotics and over the counter (OTC) decongestants have been widely used to treat chronic sinusitis. **In most cases, antibiotics are not effective for this condition because they target bacteria, not fungi.** The (OTC) drugs may offer some symptomatic relief, but they have no effect on the true inflammation.

Fungal spores become airborne like pollen. The personal air purifier is one surprising strategy of attacking the root cause of America's #1 chronic disease, by getting rid of the mold in the breathing zone of the sensitive individual. That's more than just a bonus.

That would spell S.I.N.U.S. R.E.L.I.E.F. for many. And for others, it might just spell something else. Could a personal air purifier have something to do with your mood ... ?

HOPE FOR S.A.D. PEOPLE

Do you often get depressed in the winter time? Do you know anyone who gets really down, but always during the long, dark cold season? You (or they) may be suffering from a condition called **Seasonal Affective Disorder (S.A.D.).**

This is a condition now recognized as a characteristic affective disorder with defined diagnostic criteria under the *Diagnostic and Statistical Manual of Mental Disorders* (DSM-111-R) of the American Psychiatric Association. It is not a trivial condition for it affects millions of people who literally suffer and struggle through the dark, dismal days of winter. The National Institutes of Health (NIH) estimate that each winter, 6 out of every 100 Americans suffer from debilitating depression and lethargy, as well as noticeable weight gain, as a result of this disorder.

It is known that the shortened days and much reduced periods of light exposure contribute to these depressive emotions and their consequences. Most studies and attempts to treat S.A.D. victims have used bright light presented in the morning, evening, or both, with generally strong results exceeding those with use of dim or brief light controls.

Some drug therapy has been attempted and studied, focusing on specific serotonin reuptake inhibitors (known as SRI's). There has been some limited benefit from these drug interventions, beyond placebo effects.

Now comes a surprise. Two researchers from The Department of Psychiatry at Columbia University and New York State Psychiatric Institute performed a small double-blind, randomized controlled trial (RCT) to evaluate the antidepressant effect of negative ions in the ambient air as a potential treatment modality for S.A.D.(16)

You will observe that this is the first mention in this book of any possible physiological or therapeutic value assigned to *negative* ions. This is not by neglect or default, but by design. The popular literature abounds with persistently reported salutary effects of negative ion exposure on human ailments (such as fatigability, irritability, sleep disturbance and infectious diseases). The suggestion has even been made that the worsening of such symptoms can be attributed to the opposite *positive* ion exposure, which could vary geographically, seasonally, with the weather, and in deficient indoor environments.

As a scientist and physician, this author is slow to accept the conjecture and speculation that surrounds such a controversial issue and although it may have some valid merit even in physiological terms, it seems that we should be more patient and prudent, and await more clinical research in this field before we go off on a tangent in search of unjustified claims for the *therapeutic* implications of negative ionization in the air environment.

However, it would be unfortunate if we miss the real good scientific evidence for the value of unipolar ion emission as a physical modality in cleaning up the immediate air environment. This it does by the direct energy from corona discharge; by using low levels of ozone as a chemical oxidizing agent, and by applying ionization as an electrostatic shield and precipitator in the immediate surrounding air.

Nevertheless, the results of this RCT, demonstrating that a high-density negative ionizer appears to act as a specific antidepressant for patients with S.A.D. is not only interesting and provocative, but it is also hopeful. Let's look for a minute at the evidence.

The researchers identified 25 subjects diagnosed with S.A.D. (22 women, 3 men reflecting the preponderance of females among patients coming forward to seek treatment for S.A.D.). They were properly screened and randomly assigned to two groups which received either high or 'low' ion density exposure for 20 days in 30 minute morning sessions. Other subjects were simultaneously randomized into bright light conditions. The high density (2.7×10^6 ions/cm^3) subjects showed much better response than the low density subjects. **(Again, note the intensity of ion emission as a critical factor!)**.

Interestingly (perhaps only for the curious reader), atypical symptoms (including hypersomnia, hyperphagia, fatigability and associated severe neurovegetative symptoms characteristic of S.A.D.) were affected even more by the intervention than the classic symptoms of major depressive disorder with melancholic features. There were no significant side effects reported.

The observed response rate for the high ion density exposure patients was in the range of 60%, which is also typical of the response to antidepressant medication, and not unlike that found for morning light treatment at 10,000 lux for 30 minutes.

In a follow-up study, the same authors compared light therapy to treatment with high-and low-density negative ion emitters. They found that light therapy patients did 30% better than those who got the low-density ions. But patients who got the high-density ions did just as well as light-therapy patients, indicating that the high dose ions might also be useful for treating S.A.D., with or without other adjuvants such as light therapy or medication.

Nevertheless, the mechanism of antidepressant action of the negative ions remains unknown. An ion-related serotonergic effect is only one possible speculative explanation. Others might include DC electrical fields which were simultaneously generated; oxidant gases (ozone) might be implicated; the precipitation of airborne pollutants from circulation could have indirect salutary effects. The authors even allow for uncontrolled variables such as room size, humidity, weather conditions, ambient positive ion concentrations, proximity of ground devices and even clothing and furniture fabrics which could render the treatment ineffective in a single case and blur 'between - group' differences in dose-response studies.

The bottom line here is that respected professionals have observed and published credible results to demonstrate at least the additional psychological value to S.A.D. patients of unipolar ions in the breathing zone, at ion density levels comparable to the use of the same **personal air purifier** we have been discussing.

That may not offer any serious answers to the uncertain physiologic or therapeutic value of negative ionization, but it does at least point a way forward to further research areas. It does at least offer some degree of hope ... **That's hope for S.A.D. people!**

At a time when indoor air pollution is such a serious environmental issue, with contaminants you really can't avoid and possible health consequences you certainly can't afford, you should do everything you can to protect your PERSONAL space.

"If you can't breathe purified air, nothing else matters!"

After you consider your personal space, immediately you must be thinking of the space surrounding you at HOME - that's where you spend so much of your time. That needs to be addressed …next!

TWELVE

Maintain Your HOME Space

"Your Home is all yours to Enjoy in Comfort, Safety ***and Good Health!'***

Next to your own personal space, the home in which you live is the most intimate environment you will experience. That's where you'll spend most of your life. A national survey conducted by the American Lung Association in 1999, revealed that Americans spend on average 65% of their time at home. And it is in those very homes that a different U.S. EPA study indicated that indoor air pollution may be two to five times higher - and occasionally more than 100 times higher - than outdoor levels. It's no wonder then that the same **EPA has identified Indoor Air Pollution as one of the Top 5 urgent environmental risks to public health**. That was a reminder, just in case.

That observation should also make you stop and re-examine what's going on in your individual home. What may be inherent in the design, construction and furnishings that contribute to a potential indoor air problem? What lifestyle patterns and activities may be producing contaminants that could be avoided? How could you maintain a healthier home environment?

The good news is that steps can be taken to significantly reduce, and even eliminate, many causes of indoor air pollution. We described that reduction process clearly in Chapters 3 and 6 especially. Even so, what you cannot avoid, you can dilute or replace by adequate *ventilation*. That too has its limitations. Then you can *filter* many particulates and bioaerosols that may be airborne and destined for someone's lungs. Combine the mechanical, adsorption and electronic filters and that still would not do the complete trick.

Air purification must therefore be the order of the day. Conventional methods call for ozonation, ionization and UV irradiation. In today's marketplace, real solutions are available for optimizing the quality of air in your home. Since the environmental air quality is so important, nothing less will do than the

Ultimate Solution made possible by the **FIVE IMPORTANT FEATURES** derived from the original breakthrough that NASA uses to solve the air problem inside the space capsule.

In a nutshell, if you will maintain a healthy home and avoid the consequences of poor air quality there, you must employ the **FIVE ESSENTIAL STRATEGIES** that were already discussed in more detail in Part Two.

Let's describe the Application of each strategy in turn, to the sphere of activity that comes so close to you as to merit the singular designation of 'home'. Remember, as we detail a number of practical suggestions, that '*no chain is stronger than its weakest link.*' It is all the strategies working together that will provide your ideal, healthy and safe home environment for you to enjoy.

Strategy #1. ELIMINATE IN-HOME SOURCES OF AIR POLLUTION

The American Lung Association and the U.S. Environmental Protection Agency are both very strong advocates for this first strategy. Obviously, by a process of education, each of these institutions propose to encourage Americans to take responsibility for the quality of their indoor air, especially in their own homes. **This cannot be legislated but it can be encouraged** by a variety of education programs and promotions. Naturally, the challenge is in getting the word out. With that in mind, these agencies have prepared booklets and website information postings designed to raise the standards for healthier home environments. They have also engaged in training programs for consumers, builders and home inspectors. Demonstration homes have even been built and furnished with an emphasis on improving air quality and with special considerations for both the indoor and outdoor environments.

All homeowners should educate themselves and take steps to protect and preserve their home environments which can have such serious impact on the health of their families. As the first strategy of protecting your home space, you could aim to reduce or even eliminate as many known sources of indoor air pollution as you can. It might be useful for you to go back even now and just review the various practical suggestions and ideas presented earlier in Chapters 3 and 6. Now you will have a better appreciation for what was being proposed there. To further that end, the next two pages show more exemplary tips to make you increasingly aware of the need and the opportunity. Implementing most of these steps should get you started in maintaining a healthy home environment. After all, both charity and responsibility begin at home.

SOME TIPS FOR
REDUCING POLLUTANT EXPOSURE

SOURCES

Moisture
This encourages biological pollutants including mold and other allergens. Eliminate all obvious sources. Install and use exhaust fans, especially in bathrooms. Be alert to condensation on walls, standing water anywhere and any plumbing or sewage leaks. Check roof gutters and down-spouts. Do not water soil, plants or anything else too close to the house foundation. Grade soil away from the house. Use waterproofing sealants. Use a dehumidifier, if necessary.

Carpets
Dust mites love these. Clean and vacuum regularly with a HEPA system. Deep clean at least once a year. De-gas new carpets at source or evacuate your house for a few days after any new installation.

Bedding
Change linens frequently. Use mattress covers. Vacuum the mattresses and box springs routinely.

Pets
They leave allergens everywhere. Keep pets outdoors as much as possible, if you really are a pet lover and must have your own. Limit their access to carpeted areas and bedrooms. Clean your entire house regularly, especially the 'pet areas'.

Clothes Dryer
Produces excessive moisture and dust. Regularly dispose of lint carefully from filter, and from areas around and under the dryer. Exhaust directly to outside. Use outside dry line for clothes, if possible.

Furnishings & Construction Materials
Beware of formaldehyde. Clean and dry water damaged materials immediately. Use C.R.I or GREENGUARD™ - certified materials. Following new product installations, open windows and doors and use window fans for a few days. Select solid wood as much as possible.

Draperies & Upholstered Furniture
Watch for formaldehyde - based finish. Air them out, outside. Vacuum or clean often to avoid accumulation of dust and other pollutants.

Household Cleaners
They release unhealthy or irritating vapors (VOC's). Select non-aerosols and non-toxic products with minimal chemical solvent and odor release. Follow directions. Never mix. Buy only for your immediate or near term use.

Personal Care Products
These also contain many VOC's. Select fragrance-free, non-aerosols with minimal chemical solvent release.

Paints, solvents
They usually release harmful chemical vapors. Store in a well-ventilated place, preferably a stand-alone shed, outside your house or garage.

Pesticides, Herbicides
There is always potential exposure to carcinogens, especially by children and pets. DO NOT USE INDOORS. Employ professionals only. Evacuate the house as required. Protect food utensils, children and pets. Remove shoes.

Dry-cleaned Clothes
They release VOC's. Air them out outside first before taking them back inside your house.

Car & Small Engine Exhaust
These are potential sources of carbon monoxide and combustion pollutants. Never leave vehicles, lawn movers, snowmobiles, snow blowers running in the garage (even with an open door). Seal off house from the garage, period.

Second-hand Smoke
This too is lethal. Do not smoke in your home or permit anyone else to do so!

Gas-burning Appliances
Back drafting could cause toxic gases to re-enter house. Have total HVAC professional inspection annually. Check chimneys regularly. Use CO detectors.

Fireplaces
Risk of carbon monoxide, particles and VOC's. Open flue when building a fire. Have regular professional inspection. Limit use.

Radon
Generally where present, this can cause cancer. Test your home. That's the only way to know if you are exposed. Get professional help, if necessary.

These basic tips represent just a few initial pointers to get you started. They are not meant to be exhaustive by any means. As you learn more about your particular home and more about the sources of pollution and other factors which contribute to indoor air quality, do continue to be vigilant, proactive and responsive. Strive to maintain the kind of lifestyle and engage only in those activities that enhance the quality of your environment and so safeguard your family's health in the years to come.

Strategy #2. VENTILATE YOUR HOME SPACE

We have emphasized throughout this book the tendency in modern times to design and build homes that essentially conserve energy at the expense of air quality. The goal has been to keep the cold air out in winter and the hot air out in the summer. But that meant locking up the inhabitants in a virtual air prison, confined to breathe recycled air most of the time. Without adequate opportunity to dilute or replace that air inside the home, we may have caused contaminants to accumulate and posed a real health threat that we could otherwise avoid.

The second strategy therefore among the **FIVE ESSENTIAL STRATEGIES** to purify indoor air, calls for adequate ventilation. This is a clear choice in your own home. Even if the design is 'tight' and air exchange by infiltration is restricted, there is always the opportunity to open up windows and doors as often as necessary to permit outdoor air to flow in and change the air quality dynamic. Practically, this means that it should become a pattern or regular habit to occasionally open up the place and let some (hopefully) fresh air blow through. As far as the air supply is concerned, the engineers can build you a prison for a home, but only you can make it one.

Mechanical ventilation is best used for pollutants resulting from human *metabolism*, such as carbon dioxide, methane, water vapor or ammonia. It is also essential for those activities which generate *moisture*. These might include bathing, doing laundry and dishwashing, primarily.

But most homes will need two types of ventilation. *Local* ventilation should be used intermittently to reduce local elevated humidity levels quickly, particularly in kitchens and bathrooms. Exhaust fans are generally built in for this purpose and each occupant must appreciate their value and hence practice consistent use as a matter of routine discipline. On the other hand, *general*

Engineers can build you a prison for a home, but only you can make it one.

ventilation through an efficient HVAC system should be ongoing in order to circulate and exchange the air in the entire house.

For completeness, we should mention too that certain activities, especially hobbies, tend to generate their own pollutants and need to be dealt with accordingly. Sometimes, if you are aware, less-toxic materials can and should be substituted, e.g. the artist could switch from oil-based paints to water colors, or the craftsman might use only solid woods and avoid soldering wherever possible. Nevertheless, it is often necessary to use additional local ventilation in many a hobby room. Take time to check that out.

The HVAC system in your home is not just a furnace or an air conditioner supplying hot or cold air as required to maintain a programmed temperature for your comfort. Rather, **HVAC is really an environmental system. It determines to a large extent, the quality of the air the entire family breathes as long as they are in the house.** It must therefore be given the respect it deserves. On that account, it must be professionally maintained and serviced on a regular basis.

Humidity control is also crucial. Relative humidity in houses should be maintained below 60% -- ideally, between 30-50%, depending on the climate and the time of year. Whole house humidifiers and climate control systems can help manage the moisture content of the air in your home to the appropriate level. Remember that air that is too moist provides an ideal environment for mold growth and dust mites. Neither of those you really want to have.

So to control excess humidity problems, check out the following:

- Caulk both the interior and exterior of your house, wherever necessary to keep moisture out.
- Check all windows and vents for weatherproofing.
- Fix any roof leaks ASAP.
- Direct runoff well away from the foundation.
- Check your AC coils regularly for mold growth.

You might consider purchasing an energy recovery ventilator (ERV). Operating an ERV reduces heating and cooling costs in the home while providing a continuous supply of fresh air and whole house ventilation. Be sure the system you choose can provide adequate ventilation for the occupants and help reduce contaminants.

When all is said and done, ventilation can only go so far in circulating and mixing the air, and (more importantly, in terms of air pollution *per se*) in exchanging the contaminated air *inside* with a supply of fresh air from the *outside.* This is an important strategy in its own right but as we saw much earlier, it is never enough!

Strategy #3. IMPROVE AIR FILTRATION IN YOUR HOME SPACE

It is clear that filtering polluted air is also an effective strategy for improving air quality in the home. Polluted air may contain particles of varying sizes, bioaerosols and toxic vapors and gases. All these can effectively be removed by passing the air through mechanical filters, adsorption filters or electronic air cleaners attached to the central HVAC system. This would be a start, so let's first make some general comments.

- Install air conditioning so you don't have to keep opening windows and doors to get fresh air. As you keep those windows and doors closed and the AC on, you will help to prevent pollen and other outdoor allergens (and other undesirable pollutants) from entering the home.
- Furnace and air conditioning filters should be changed regularly every two or three months, or otherwise, as required by the manufacturer. Additional HEPA filters are recommended for people with particular sensitivities to dust.
- Dust can be a major irritant indeed. Therefore, dry mopping or dusting, which merely scatters dust into the air and (hopefully) to be trapped by the filter, should be avoided. Use a damp cloth or mop, in preference.
- Older homes do sometimes have furnaces or pipes covered with asbestos-containing materials. If you have a suspect home like this, have a professional check it out and if the asbestos-containing insulation is falling apart, then hire a professional contractor only, to deal with it.
- If you have a ducted HVAC system, check the fresh air intake on the house exterior to make sure that the fresh air intake is unobstructed, well above ground and upstream from any potential source of local contamination.
- Forced air ductwork does require occasional maintenance and possible cleaning to make sure it does not become, by itself, a further source of indoor pollution. But before you have it cleaned, you should verify if the ducts are actually causing health problems in your family. Otherwise, it might be better to leave them alone.
- All ducts should be tightly sealed so that air does not leak and contaminants enter in this way to circulate throughout the house.
- Check that all inlets and outlets of the central system are not blocked by furniture or other large objects to impair airflow.

So much for those general comments. To be truly effective and get the job done, you must go beyond the typical all-purpose systems that are generally provided in most houses in America today. Wherever circumstances warrant, much better technology is available for local selective use that will deliver much better results, in terms of efficiently removing most of the airborne contaminants as desired.

There are a number of commercially available free-standing HEPA air purifiers on the market today. These models use pleated *High Efficiency Particle Arresting* filters to trap airborne allergens quite effectively. Recall that HEPA filters can remove 99.97% of particles larger than 0.3 microns, certainly enough to help improve indoor air quality and reduce exposure to irritants and allergens. HEPA air purifiers use a fan to pull air through the filter of the free-standing unit and naturally, only the air that actually passes through the filter gets treated in this way. They produce no ozone at all (which could in reality be a disadvantage, considering the proven benefits of low-level ozone within the safety limits). Some are quieter than others, but the fans tend to be noisy in confined spaces.

Of course, HEPA filters are available, but there's much more to the current technology. **Cartridge filters**, for example, are available in various designs for different specialized purposes. Ultra pleated particulate cartridges greatly expand the surface area available to trap suspended particulates from the air. Just to illustrate, a 66 square inch ultra pleated cartridge filter might have an effective surface area of about 50 times that nominal value (as much as 3240 square inches) and equivalent to a nominal HVAC filter which would have to be a jumbo 4.5' x 5' in size. (That alternative is ridiculous!)

A cartridge filter for removing any potential *mercury* from the air would consist of a specialty bed of sublimed sulfur/carbon mixture designed for maximum adsorption. A typical cartridge could adsorb as much as 148 gm of mercury. Similarly, a cartridge to trap VOC's might again consist of a blend of activated carbon and natural elements for adsorption of a wide variety of VOC's. In this case, a typical cartridge could adsorb up to 190gm of volatile organics.

Radon is a special case. A cartridge is available that consists of a blend of natural elements that are polyactivated for efficient adsorption of radioactive radon gas. This typical cartridge could adsorb about 225gm of radon.

The point is that specific filters are available commercially today, often incorporated in a more complete air purification device that can get the job done when special circumstances demand.

Strategy #4. PURIFY THE AIR IN YOUR HOME SPACE

Caveat Emptor (Buyer beware!)

There is so much senseless debate about the relative effectiveness of different strategies for air purification. The fact is that no single strategy is either perfect or complete. **The solution to the problem of indoor air pollu-**

tion requires a *comprehensive approach* involving a number of different strategies. That's indispensable.

So, while you take all the effective measures to *reduce sources* of air contaminants in your home; while you *increase ventilation* to dilute and replace the indoor air in a way consistent with comfort and cost-efficiency, and as you take steps to install and maintain *appropriate filters* to reduce suspended particles, bioaerosols and volatile gases and vapors - you must also address the limitations of all those three strategies and take measures to purify the air directly. As we said earlier, **in war there is defense** and that is very important - **but battles are generally won on offense.** In war, you go after the enemy with all you've got and fight to the death.

However, in this area of air purification, be wary of old technology! Be jealous of scams! The field of indoor air quality investigation and remediation is relatively new. As a result, many opportunists have emerged to profit from your assumed probable lack of knowledge and your innate desire for a healthy environment. **They will claim what they cannot prove and promise what they cannot deliver.** But in the space age, one must deliver and be able to demonstrate that they do!

Conventional Air Purifiers

Here you have choice, too much choice. The key is to focus on the technology and its application. **Conventional air purifiers** will fall into one of the three broad classifications: ozone generators, ion generators or germicidal irradiators (lamps). Each of these technologies has inherent limitations and disadvantages of which you must be aware.

Be wary of old technology! Be jealous of scams!

Ozonators can, in principle, produce too much ozone. Federal guidelines today limit exposure to 0.05 ppm. Above this concentration you will be at risk of experiencing the irritating and even potentially toxic effects of ozone. *Ozone devices must therefore have the inherent technology or built-in controls to limit ozone production in ambient air during normal use.* The smell of ozone is a convenient detector to guide you in common application. Most people can detect ozone down to about 0.01 ppm and at that level, the aroma is pleasant and fresh, reminiscent of being in the mountains or near the ocean - the kind of scent you experience outdoors after a thunderstorm. But at or above 0.05 ppm, the odor is more pungent and irritating so that the warning bells should alarm any normal person to turn-it-off and evacuate immediately.

For low-level ozone in residential application, UV-C ozone lamps are frequently used. *Corona discharge* ozone units may find commercial application but great care should be taken with residential applications in occupied areas. If improperly used, they can exceed federal safety limits of ozone. When choosing either a UV-C ozone lamp type air purifier or a corona discharge ozone air purifier, look for a system which provides in any case, for adjusting the output to match the ventilated space.

In some cases, if your house is going to be evacuated for normal reasons, the device may have an '*away-mode*' operation setting. This will increase the maximum output of ozone (perhaps enhanced by CD) for increased effectiveness in unoccupied spaces. Alternatively, a '*high mode*' or '*fast-mode*' setting might be used to deliver an intermediate range for faster coverage and increased smoke and odor elimination. All this is appropriate, as long as you use this kind of device with responsibility and discretion. The point is that ozone should never become an exposure health threat. This need never happen with today's technology! (RCI to the rescue, later!)

Ion generators use static charges to remove particles from the air. They are different from electronic air cleaners like you may have installed on your central HVAC system. These latter cleaners are usually electrostatic precipitators or charged media filters that use an electrical field to trap charged particles. In electrostatic precipitators, particles are collected on a series of flat plates. In charged-media filter devices, which are less common, the particles are collected on the fibers in a filter. Either of these may have an ionizer to charge the particles before the collection process in order to improve collector efficiency.

Ion generators on the other hand, are portable units that act by spitting out ions into the air that get attached to suspended particles. These cause the particles to become charged in such a way that they are attracted to walls, floors, table tops, draperies, occupants or anything else around. In some cases, these devices may contain an oppositely charged collector plate to attract the charged particles back to the unit.

The limitations here are obvious. Soiling, as mentioned previously, can be a major problem. Even if you resort to vacuuming, you've got a lot of work to do, and many areas will be inaccessible.

As we pointed out in the previous chapter, some manufacturers of ion generators are quick to claim that negative ions in the atmosphere are energizing to people and may have all kinds of positive health influences. But this is essentially subjective, with unimpressive science to clearly substantiate the associations. There may be good reason for further research in this area, but to date any rationale for employing ion generators as a therapeutic treatment modality only, seems unjustified.

The plain truth is that, by itself, this technology remains an ineffective practical method for improving indoor air quality. There are much better alternatives today.

Germicidal irradiation lamps have become quite popular as an easy fix for the problem of mold that tends to build up on air conditioner coils. They are generally low cost, require minimal installation and are quite effective in that regard. But they are really not a practical solution for airborne mold.

Germicidal lamps have been used centrally, inside the HVAC system, as an additional purification method. Units are available for relatively easy installation inside the ductwork. Also, as we pointed out much earlier, special lamps may be attached directly to the AC coils to attack any potential mold build up. These have varying degrees of efficacy.

For disinfecting room air, the best application has used the 'upper room' configuration where the UVGI device is mounted in the ceiling or at the top of adjacent walls. The current UV dosing guideline remains one 30-watt suspended figure or two *30-watt wall fixtures for each 19 square meters of floor area.* Under normal conditions, the added health hazard of less penetrating, low-intensity, indirect UV-C exposure, reflected from the upper room, is minimal. However, with the space age technology available today, this is certainly not, by itself, a preferred method for disinfecting room air either.

Conventional Air Purifiers often use varying combinations of the technologies we have outlined earlier - filtration, ozonation, ionization and UV-radiation. This is clearly seen in **Table 13** which identifies the different technologies used in some of the leading conventional air purifiers on the market today. They all have intrinsic limitations that can be addressed by the new space-age RCI™ technology.

Table 13 Conventional Air Purifiers' Technologies

Air Purifier	Technologies	Area (sq.ft)
Oreck Super Air 8	Prefilter Charcoal filter Electrostatic Precipitator Negative Ion Generator	900
Sharper Image Ionic Breeze	Electrostatic Precipitator Germicidal UV Negative Ion Generator	400
Ionic Breeze Quadron	Electrostatic Precipitator	350
Friedrich C-90A	Prefilter Activated carbon Electrostatic Precipitator	600
IQ Air Health Pro	HyperHEPA filter	900
Blueair 601	HEPA filter Activated carbon	679
Whirlpool Whispure 45030	Prefilter HEPA filter	500
Bionaire BAP 1300	HEPA filter Prefilter Activated charcoal	488
Honeywell HEPA 50250	Activated Carbon prefilter HEPA filter	374
Hunter QuestFlo 30400	Activated Carbon prefilter HEPA filter	400
Vernado AQ 535	HEPA filter Activated carbon	400
Hunter HEPAtech 30375	Prefilter HEPA filter Ionizer	400

The Choice is Yours

The science of air purification is quite complicated and understandably confusing. Many consumers (and their organizations too!) are at a loss to interpret the information that is available in this field. In the residential market for these devices, millions of units have been sold, with quite a range of success in their application. Since the events of 9/11, there have been ongoing concerns about environmental terrorism and also widespread television coverage of infectious disease outbreaks like SARS. Consumers wish to protect themselves from such potential threats, for sure, but heightened awareness has increased demand for methods of cleaning up the air we normally breathe on a daily basis. The contaminants of normal air are becoming more widely recognized and informed consumers want to take nothing for granted.

But aware consumers should obviously take the time to gather reliable information about indoor air quality and to explore the technologies available today to deal with the problems. You must evaluate the nature, level and sources of contaminants that are likely to be important in your particular home space.

However, remember as you do, that *it's not just what you can see that really threatens you, it's also what you cannot see.* Then, when choosing an air purifier to meet your specific needs at home, it is important to understand what the device will do in practical terms to remove any particles, bioaerosols, odors, vapors and gases from your surrounding air. The technology is all- important for it will determine the unit efficiency and your air quality experience. Of course, you must also weigh initial costs, maintenance costs and power requirements. But remember, you will only get what you pay for. Something that's cheap, usually is just that ... cheap! Consider also the size and layout of your home. What kind of coverage will you require? These are the kinds of things that you must factor into the equation. Again, you will, after all, only get what you pay for, and perhaps not even that much - unless you carefully discriminate.

However, in the end, all the conventional air purifiers that we have mentioned so far, will fall short of the space-age technology that we described earlier - Radiant Catalytic Ionization™. It is the synergy of all FIVE IMPORTANT FEATURES that will provide your ultimate solution! That's your final strategy ...

Yes! There is a more excellent way!

Strategy #5. **USE SPACE-AGE TECHNOLOGY AT HOME**

The latest Radiant Catalytic Ionization™ (RCI™) technology combines the best of all the conventional air purification technologies with potent synergy.

Fundamentally, as we have described earlier, RCI™ uses the photoelectric effect of photocatalytic oxidation (PCO) to generate powerful radicals that react with pollutants in the air as it is drawn through the cell/probe. That accounts for the traditional effects of *ozonation* (comparable to natural air cleansing outdoors) and much, much more. Secondly, it produces ions that leave the unit as a purifying plasma that is effective in the room air wherever the ions encounter pollutants. The commercial units might also include ion generators (both needle point and radiofrequency types) that produce negative and positive ions to provide all the benefits of *ionization*. Thirdly, RCI™ uses a high intensity UV lamp with germicidal action on microbes passing through the unit - hence, the benefits of UV-germicidal *irradiation* (UVGI). All three major air purification methods are combined into one, with remarkable synergy. That's how nature does it outside. In addition, the very low levels of ozone (0.01-0.02 ppm) leaving the unit, make all the controversy and paranoia about unhealthy effects of ozone, both untenable and irrelevant. **Ozone exposure is simply no longer an issue under any normal circumstances in which RCI™ technology is used.**

Ozone expousre

is simply

no longer

an issue, with

today's technology

In practical terms, RCI™ is used in two different modalities in the home space. First, as a free standing unit (console), it can be strategically placed in your home to effectively cover a range of floor spaces, anywhere from 250 square feet up to 3000 square feet. The effective coverage will naturally depend on different variables such as home design, severity and frequency of pollution, humidity and temperature. Alternatively, it can be mounted directly as a probe/cell inside the ***Duct Work*** of your central HVAC system. In the latter case, the HVAC blower fan can amplify the purifying effects of the RCI probe, by distributing the purified air and plasma from the basement to the attic, and everywhere in between.

It is worth noting again, just for emphasis, that because RCI provides *active* air purification - **taking the solution to the pollution** - the probe is placed not in the return duct to clean up the returning air, but in the duct *after* the blower fan, to allow the purifying plasma to be distributed throughout the house.

That is an essential difference!

So, after your best attempts to reduce known pollutants at their source, and after the best ventilation practices have been implemented in your home space, to complete the **FIVE ESSENTIAL STRATEGIES**, you will need to employ an RCI probe/cell, either in a portable unit or installed in your central HVAC system. And don't forget the need for filters - preferably a combination of filters - especially a HEPA mechanical filter for most particulates, allergens, aerosols and microbes, and a carbon-type adsorption filter for vapors and gases, that also include volatile organic chemicals and odors. A cartridge combination design will probably be best.

RCI ™ technology takes the solution to the pollution throughout your home!

Since much of the air purification takes place near the catalytic surface as the air is circulated through the RCI™ cell/probe, and since the purifying plasma leaving the cell is so effective, a variable speed fan (preferably, very quiet) will go a long way to increase efficiency and increase the CADR (clean air delivery rate).

When employing a free standing (console type) unit, you should place the unit as high as possible off the floor, to ensure complete purification. Putting it on a bookshelf or cabinet, for example, you should also place the unit either nearest the source of worst pollution (if that is apparent); *or* secondly, near a cold air return (to thoroughly circulate clean air throughout your house); *or* thirdly, in the area most heavily used (to achieve maximum benefit). The rear of the unit should always have some open clearance (at least an inch or so) to allow unrestricted airflow.

With the application of this space-certified RCI™ technology in your home, you will enter the space age. After all, your home should be your castle, and your castle should be permeated with that clean, fresh scent of nature, reminiscent of being in the mountains or beside the ocean. An inviting, healthy indoor environment is yours to discover and yours to enjoy. You can make the *application* and capitalize on this 21st Century *solution* to a late 20th Century *problem*.

Enjoy your Healthy Home! Enjoy your Fresh Air for Life!

Note: FYI- *As we described in the last chapter, the space-age RCI™ technology was originally developed by the RGF Environmental Group out of Florida. The application (portfolio) of this technology use was purchased in the summer of 2004 by EcoQuest International, with headquarters in Greeneville,*

TN. These two companies also have a long-term strategic alliance going forward, whereby RGF will continue to provide EcoQuest with ongoing Research and Development as well as technical support and training, in addition to further developments of intellectual property. EcoQuest therefore owns the trademark and the exclusive right to the RCI™ *technology and continues to expand both development and application of Radiant Catalytic Ionization™ as well as its worldwide distribution.*

THIRTEEN

Protect Your TRAVELING Space

"If you can't control what other travelers do, at least ***control the air you breathe"***

We've looked at defending your personal space - the environment that you inhabit 24/7. It is small, private and constantly about you. Then we examined the challenges of maintaining your home space. That is a little larger, it's where you spend most of your time and hopefully, where you derive most of your joys. You share it with others, but they are close to you. They hopefully share your values and respect your standards. You can appeal to their best and loyal instincts for their cooperation and respect. Together you can practice the mantra that **Health is Job #1**.

But you cannot remain in the shelter of home forever. Sooner or later, you must get out from that protected environment into the real world where people care less and practice lifestyles all their own. Not everyone will be aware and certainly not every one will buy in to the conviction that **The Air We Breathe is Necessity #1**. You now know that the *quality* of the air around you is therefore of paramount importance. You also know that you can promote it to your good health whereas if you pollute or just neglect it, you would do so to your own detriment. But that's not a common concern or priority for everyone else around you.

So, as you travel away from home, you must share with your fellow travelers, the same air space for which they just might not care. At least, not the way you do now. But be that as it may, travel you must. You can at least take comfort in the fact that you will not be alone.

North Americans are a forever traveling public. With absolute freedom of movement, a well-developed infrastructure of roads, rails and airports, an abundant supply of automobiles, trains and commuter flights, and the means to afford the energy costs - there is no limit to either where, when or how we go from place to place. You get the picture. We are constantly on the move.

This ease of transportation and mobility helps to drive the largest and most efficient economic engine this world has ever known. But we do so at a price. First, we exploit mechanical transportation at the expense of personal exercise and that combined with our appetite indulgence and high calorie, fast and convenience food culture, has made obesity the epidemic it has become in recent times. That's an important consideration for another time and place. However, it is the other important implication that concerns us here. Namely, that **we travel in cars and planes especially, in an indoor environment that puts us all at risk.**

The International Center for Technology Assessment (CTA) is a non-profit corporation, formed in 1994 to assist the general public and policymakers in better understanding how technology affects society. They have issued a series of reports designed to assess the environmental and social impact of transportation technology. These reports were designed to aid both the public and the policymakers in their ongoing deliberations concerning the future course of transportation in the United States.

They came to some amazing conclusions. In their fourth report, they made an assessment of the air quality inside automobile passenger compartments and the title of the report said it all:

IN-CAR AIR POLLUTION
The Hidden Threat to Automobile Drivers

This report(1) found that pollution levels inside cars are often much higher than those detected in the ambient air, at the roadside and in other commonly used public transportation vehicles. In that review, they found that the results of 23 separate scientific studies conducted during the 1980's and 1990's reveal that in-car air pollution levels frequently reach concentrations that may threaten human health. That's exactly what the data shows. The air inside cars typically contains more carbon monoxide, benzene, toluene, fine particulate matter and nitrogen oxides, than the ambient air at nearby monitoring stations used to calculate government air-quality statistics.

This is therefore a major health and environmental problem. Americans are driving more than ever. We choose to drive almost everywhere - to work, to the supermarket, to family vacation spots, to visit ... to who knows where. In aggregate numbers, we traveled more than 2.8 trillion miles (that's 2,800,000,000,000 miles!) by automobile in 1995, up half a trillion in five years and double the number of thirty years earlier. We therefore spend more time than ever before, inside cars. With increased traffic congestion, we spend more time going shorter distances. Many people commute to work on a daily basis, facing a drive time of about half an hour or more each way.

But we are being deceived. Automobile manufacturers have done everything to make the comfort and apparent safety of modern cars appealing to our consumer appetites. Just think of stereos, heated leather seats, computerized, dual-zone climate controls, power seats with memory, and the list goes on. With such common luxuries, we would think that the interior of our cars was an ideal place to be. We should just sit back, relax in style and enjoy the ride. However, nothing could be further from the truth.

In effect, cars on busy roadways drive through an invisible tunnel of concentrated pollutants. These hazardous pollutants accumulate inside cars driving in moderate to heavy traffic to make the daily commute a realistic health threat that should be the focus of public alarm.

It has become part of our daily media fare to listen to reports on local weather and pollution levels. Public health officials are quick to issue warnings whenever concentrations of auto pollutants in the ambient air exceed levels that pose a danger to health. It is as if we should stay inside our homes and our automobiles to avoid the environmental danger of standing or walking outside in such toxic air. But we fool ourselves if we fail to acknowledge the fact that the air quality inside the car en route to work is typically much worse than it is at the side of the road. That's what the science tells us, no matter how privileged and comfortable we feel in our top of the line automobiles.

Particulates

The most dangerous component of automobile exhaust is in the form of particulate matter (PM). These are the same tiny particles we discussed earlier. They are fine enough to go deep into the lungs and cause havoc. Some health effects are relatively minor like nasal congestion, sinusitis, throat irritation, coughing, wheezing, shortness of breath and chest discomfort. But other cumulative effects can prove deadly. Recall that the 'Six Cities Study' by the Harvard School of Public Health found that test subjects exposed to higher PM concentrations were 26% more likely to die prematurely than subjects exposed to lower concentrations.(2) Similarly, a 1995 study from Brigham Young University found that test subjects exposed to higher PM levels were 17% more likely to die prematurely compared

The air quality inside the car en route to work is typically much worse than it is at the side of the road.

to those with lower exposure levels.(3) All in all, **tens of thousands of Americans die prematurely due to exposure to particulate matter each year**.(4)

The vehicles that use diesel fuel are especially guilty of PM exhaust. **Levels of particulates can be up to eight times higher within a car** traveling behind a diesel truck or bus compared to the PM level in the air at roadside. Europeans drive more diesel vehicles compared to North Americans, but we are not exempt from the consequences on this side of the Atlantic. What is more, over 90% of the particles found inside vehicle passenger compartments on U.S. roads measured less than one micron in diameter. Those are the more dangerous particulates that can cause serious negative health effects. The researchers who did this 1994 study found average PM levels of 105 µg/m3 (micrograms per cubic meter) during highway driving conditions, enough to conclude that "***the passenger compartment air quality can be described as unhealthful***".(5)

> *"The automobile has not merely taken over the street, it has dissolved the living tissue of the city ... Gas-filled, noisy and hazardous, our streets have become the most inhumane landscape in the world."*
>
> — James M. Fitch

The situation is worse if there are more diesel vehicles on the road, or if test cars travel *on gravel roads* with the windows down, or if they're tested in dense city traffic. Conventional attempts to filter out these PM pollutants were not very successful. Air conditioning systems removed between 40% and 75% of the *larger* particles but as far as the submicron particles were concerned, the reduction was only 2% to 15%. A typical interior air filter reduced the large particles by 90%, but again, could only remove just 5% of the smallest particles.(5) Drivers and passengers therefore remain unprotected from these more dangerous fine particles.

That includes you.

Volatiles

Both gasoline and diesel automobiles produce exhaust that contains significant concentrations of about two dozen volatile organic compounds (VOC's). Several of these chemicals are potentially harmful to the environment and to your health. They include benzene, 3- butadiene, xylenes, ethylbenzene, toluene and formaldehyde. Some of these are known carcinogens. But it is important to note that your individual risk is from your cumulative exposure. In other words, you may think nothing of a 'whiff here or a whiff there' of *this or that* chemical pollutant in your car. However, because you are probably spending more and more time driving or riding in automobiles, the fact that carcinogenic VOC levels are elevated should certainly arrest your attention.

This is a toxic contribution you cannot afford to ignore.

In-car concentrations of these VOC's are significantly higher when driving in the city, compared to driving in the suburbs or on the highway. It has also been shown that improperly maintained vehicles tend to have significantly higher in-car VOC concentrations than well-maintained vehicles. The VOC levels are slightly lower when cars are driven with their air conditioning system turned on. The peak levels are found during commuter trips marked by heavy traffic or frequent stops at traffic lights behind other vehicles. **The in-car concentration levels typically go up to 14 times as high as the maximum seen at the roadside.**(6) They are about ten times higher than levels measured inside commuter trains making similar trips.

To put this in context, just think of benzene. This is a known carcinogen, so much so that the World Health Organization (WHO) set the acceptable human exposure level for benzene at zero. But Harvard researchers found that drivers were breathing in more than one-fifth of the total amount of benzene they inhaled over the course of an entire day, during the one hour and fifteen minutes they spent in their cars.(7)

The 1998 California Resources Board study measured 13 different VOC's inside a pair of cars on a variety of roads in Los Angeles and Sacramento. They also measured corresponding levels immediately outside the cars in the traffic stream, and at the roadside. The results as you might expect again confirmed that the VOC concentrations are higher inside the car, compared to levels at the roadside or in ambient air measured at remote fixed sites. The in-car concentrations of all the VOC's were generally similar to those measured in the traffic stream immediately outside the test vehicles.(8)

Next time *you* are stuck in traffic … pause to think of benzene, xylene, toluene, formaldehyde and all the other forgotten organic pollutants that you might be inhaling. Think of these molecules as an invading army taking aim at your tiny lung alveoli, breaking through their fragile walls and going out with your blood circulation to do their dirty work.

That should again be another wake-up call.

Carbon Monoxide

No discussion of air pollution inside vehicles could be complete without the mention of carbon monoxide (CO). This is a ubiquitous gaseous byproduct arising from the incomplete combustion of hydrocarbon fuels, including gasoline and diesel. A late model, properly maintained car will emit about 420 lbs of CO each year, compared to 547lbs for an SUV. Emissions of CO are higher from cold engines, in colder weather, at lower speeds and in city traffic.

Each year in America, some 10,000 people seek medical attention or are incapacitated due to carbon monoxide poisoning. About 1500 people commit suicide by intentional CO exposure from confined car exhaust. Another 1500 die from unintentional, automobile-related CO poisoning annually. The National Highway Traffic Safety Administration found that in 1993 nearly **one third of the accidental CO fatal poisonings that resulted from auto exhaust, involved drivers or passengers in moving vehicles.**(9)

Research shows that CO concentrations in cars consistently measure higher than those at the roadside or inside other types of commuter transport vehicles (buses, trains). More than twenty years ago, the U.S. EPA and Rodney Allen of Comp-Aid, Inc., measured CO concentrations inside test cars on commuter routes in Los Angeles. They found that they were nearly identical to those immediately outside the vehicles and were, on average, nearly four times as high as levels recorded at remote monitoring stations. They were highest in heavy, stop-and-go traffic conditions and results showed that opening or closing the cars' vents had no significant effect.(10)

When it comes to CO in particular, a bad situation would be to find yourself in dense, stop-and-go traffic in the winter time, trapped behind high-polluting vehicles, such as an out-of-tune delivery truck or an older car lacking emission control systems. That really would be bad news all around! You'd be in a soup and with no way out. That is, unless ... well, you'll have to just wait and see.

There's one other thing to compound that scenario even more. You guessed it. ***Smoking*** **by automobile passengers had a noticeable effect on in-car CO levels.**(11) They often exceed 100ppm if somebody smokes, whereas average interior CO concentrations during non-peak traffic times range from only 10-25ppm. In rush hour, non smoking commuters typically have to contend with CO concentrations of about 30-40ppm. For comparison, average ambient air CO levels from measurements at fixed stations near commuter routes are typically less than 3 ppm.

Because carbon monoxide has no color, no odor and no taste, and because moderate exposure may produce nothing more than flu-like symptoms in healthy people, many individuals who suffer from non-fatal CO poisoning probably remain unaware that they have been exposed to the gas. The vast majority of cases most likely go unreported and untreated. Now that you do know what you know, you will have to maintain a heightened sense of suspicion whenever you feel less than your best, after a long drive in dense traffic.

SOLUTIONS

Elevated in-car pollution levels particularly endanger children, the elderly and people with asthma and other respiratory conditions. But that is in an acute sense. All of us, including even the healthiest among us, must still be concerned about the cumulative effects of such exposure. In-car air pollution has indeed received little media attention - that's true - but that does not alter the fact that it poses '***one of the greatest modern threats to human health***'.[(1)]

That's the major conclusion of the CTA report on *Indoor Air Pollution - the hidden threat to automobile drivers*. Their policy recommendations called for increased funding for public transportation, pressure on automakers to develop zero-emission vehicle alternatives and new emission regulations. Regarding individual responsibility they would only suggest that you drive less, use more public transport and when you do drive, make sure you properly maintain your vehicle to reduce its emission and maintain auto body integrity.

What else can you do? Here are ten common sense tips to reduce your risk:

TEN COMMON SENSE TIPS FOR DRIVERS

- Impose a smoking ban when you drive.
- Scatter your work hours, if possible, to avoid rush hours.
- Avoid driving behind old cars and trucks.
- Keep a fair distance behind - tailgating is dangerous!
- Don't idle the engine in parking lots or garages.
- Use less congested car pool lanes with no trucks.
- Close vents (hit the recycle button) if smoke or dust is visible.
- Avoid air fresheners/deodorizers/chemical cleaners in your car interior.
- Drive when you must, but walk, run or bicycle when you can.
- Use a personal air purifier and space-age technology for **Fresh Air as you go.**

But that is still not enough. You have two additional options available.

The first is to use a **personal air purifier** that you can proudly wear around your neck, just as we described in Chapter 11. This simple application will allow you to benefit from the high intensity ion streams that will impart charge on to the airborne particles and help to remove them from the air. Remember that it is the smallest particle fraction that proves to be most intractable and the most dangerous to health. Remember too, that peer-reviewed research from the University of Cincinnati and also from the University of California, (reported in Chapter 11) has shown that fine and ultra-fine particles are effectively removed by this space-age ionization technology. That's the target here with your personal air purifier. Put it to work for you!

The second option available is to use the space-age RCI technology described in Chapters 10 and 12. Commercial units are available specifically for this application. They literally provide **Fresh Air as you go**. They are battery operated, can be plugged into your cigarette lighter, or mounted directly on your dashboard. Within seconds, the power of this technology can go to work inside your vehicle as you commute to work, take a family vacation or take a trip anywhere else. It will eliminate smoke and odors and a broad spectrum of airborne contaminants in the confined space of your automobile where you spend an inordinate amount of time.

In the modern world, traveling *by car* is hardly an option. It is almost a daily necessity. And not infrequently, traveling *by air* is also an unavoidable inconvenience and chore (if the truth be told). Walk through a busy airport in any American city and you would think that the whole world was flying somewhere. We're now given to commuting by plane, but especially since 9/11, we are constantly obliged to think of its attendant risks. However, there is one risk or threat that we often overlook. That makes it no less real.

IN-PLANE AIR POLLUTION
The Other Threat to Frequent Flyers

Airplanes and airports are known to be major sources of noise, water and air pollution. Living near a major airport is a constant reminder of the price of urbanization and industrialization. The sound of aircraft taking off and landing is at times an irritating distraction, but that is probably the least of the environmental impact. Jet engines constantly pump carbon dioxide, volatile organic compounds and oxides of nitrogen into the atmosphere. Consider also the toxic chemicals used to de-ice airplanes during winter, which can end up in the waterways. This type of pollution is obvious and gets environmentalists all worked up. But despite their best efforts, air traffic (with the attendant pollutants) is on a rapid increase worldwide.

What escapes most observers is the other type of pollution that affects the individual air traveler sitting in relative comfort inside these flying machines. Did you ever wonder about the quality of air inside the plane you were traveling on? After all, you're trapped in this small, metal container several miles up in the sky, sharing the limited space with unknown companions who have uncertain medical histories and conditions.

A few hours in a poorly ventilated airplane can be very uncomfortable for most people. For some, with chronic respiratory conditions or a weakened immune system, the risk of infection or simple exacerbation of their symptoms may prove burdensome. But in any case, the quality of interior air that travelers must breathe when flying is of major concern to all the different stakeholders in this industry. The vulnerability of the individual traveler, the potential liability of the airlines, the reputation of aircraft designers, manufacturers and maintenance crews, and the regulatory responsibility of government agencies, all coalesce on this air quality issue too - just as they do on other matters of aircraft safety and efficiency.

However, the Air Transport Association which is the trade association that represents U.S. commercial airlines, maintains a defensive position on indoor air quality. Their assistant general counsel, Katherine Andrus, was quoted recently in *The Wall Street Journal*: "the air is safe, it's clean and there should not be any need to seek any device to protect people." (12) The most they would concede is that the data from several studies on the effects of breathing recirculated air in plane cabins are at most, inconclusive. You may indeed get sick from air travel, but they would question what the cause(s) might be.

Airplane Ventilation Systems

Times have changed. Before 1970, the average passenger aircraft provided 15 cfm (cubic feet per minute) per person of outside air. When the oil crisis hit in the early 1970's, we changed not only the way we designed our homes and cars, but our airplane designers were also obliged to focus on energy conservation and among other bigger changes, they redesigned planes to recycle up to 50% of the air in order to conserve on fuel. Until then, 100% fresh air was brought inside the cabin. The last jet model made with that old design was the Boeing 727 aircraft. Since that one, more modern airplanes provide 5 to 10 cfm of outside air which is misnomered as 'fresh,' although it is generally taken from the engine intake compressors.

At normal flying altitudes, the ozone concentration in the outside air is relatively high, so the air coming in is passed through a catalytic ozone converter to remove the potential irritant. At times, in some limited aircraft like the

BAe-146, oil leakage from oil seals has been known to overload the catalytic converter and allow smoke and lubricating oil and components to enter the air of the cabin.(13) That is clearly the exception, and not a frequent occurrence, but it illustrates what can happen.

Apparently, common complaints of itchy red eyes, headaches and even disorientation have been more commonly reported among flight attendants in the cabin than among pilots. One possible explanation is that the cockpit is designed with a completely separate ventilation system that supplies much more fresh air per person than the cabin system.

The 5 to 10 cfm of outside air brought into the cabin, is supplemented with another 10 cfm of recirculated and filtered air. Filtration of air is highly variable. Some small airplanes have no filtration system whereas most large commercial aircraft use HEPA filters which you recall are supposed to remove more than 99% of particles greater than 0.3 microns. Most airborne particles as well as bacteria and fungi are larger than a micron, so they are effectively filtered out. The viruses are another matter. They are only about 0.01-0.10 microns by themselves, but they often clump together to form particles larger than a micron which could then, in principle only, be also filtered out by the HEPA filter.

With 50% outside air and 50% recirculated air in most aircraft today, the cabin air is exchanged every 4 to 6 minutes compared to every 5 to 12 minutes in typical offices and homes. But that's not the whole story. The airflow patterns in commercial aircraft are unlike those in standard indoor ventilation systems. The typical design has air inlets in the ceiling, and return through the floor. The air therefore travels across the height of the cabin and much less along its length, from the front to the back.(14) Air tends to flow side to side in pockets that are no more than a few rows wide, according to Boeing.

Then there is the question of congestion. Economics calls for a greater density of people per unit floor space or per unit air volume. A full, narrow-bodied plane fitted with the maximum number of seats may provide as little as 1.5 cubic yards of interior space per person. This is less than one-third the space in a sold-out indoor auditorium.

Surveys show that in 1999, for example, a typical flight had 93 passengers and 5 crew on board. On average, five of the passengers had a cold or flu. That means that if you were a passenger in a row of say six across, you had a better than 50:50 chance of sitting close to a sick passenger within a row of you. In an epidemiological study published in 2004 in the *Risk Analysis* journal, researchers found that the chance of catching tubercolosis on an airplane was greatest for passengers seated within two rows of each other.(15) In this environment - filter or no filter, 'fresh' outside air or recycled air, it does not matter - you would be at risk of inhaling some of whatever that poor fellow passenger was putting out into the surrounding air. Then just imagine what may be happening if you ever have to travel on a plane next to a person who is constantly coughing or sneezing. That is a scary thought.

You could clearly be at risk of infection.

Airborne Infections

In 1997, the International College of Surgeons, U.S. Section, introduced a resolution which asked that - "*The American Medical Association (AMA) encourage well-designed studies to discover the major causes of airborne infections on commercial flights, as well as reasonable solutions for minimizing this health hazard to travelers on such flights; and further, discuss the types of airborne diseases, methods of spread of these diseases and possible methods of prevention within the airline industry, the FAA and other appropriate interest groups and agencies, and strongly encourage voluntary measures to minimize such health hazards*."(16)

This resolution was referred to the Board of Trustees of the AMA because testimony at the Reference Committee indicated that the existing data did not suggest airborne infections on commercial flights was a significant problem. The speakers stressed that 'airborne infection is primarily an issue of confined spaces and less an issue of airline conditions.'

The AMA did a thorough literature search using the MEDLINE database covering 1980 - 1997, and held discussions with ASHRAE and NIOSH. They more or less came to the same conclusion that 'under usual aircraft operating procedures, cabin air quality does not represent a significant risk for transmission of airborne infections.'

They observed that most of the concerns and complaints brought forward by cabin crews and patients, in general, were non-specific symptoms hardly attributable to identifiable biological contaminants. Studies found that bacterial and fungal counts in commercial aircraft were lower than those in public buildings. The levels were very low and unlikely to cause adverse health effects. In fact, measured air quality parameters easily exceeded the requirements for maintaining a healthy indoor environment.(16)

That's all well and good but there should be at least **four qualifying observations.**

The first is that airborne *viruses* are the most pressing concern and the technology for identifying these *airborne viruses* in flight is still lacking. So viral loads remain somewhat of an unknown. Even today, one cannot rule out viral spread in aircraft as a significant problem, especially with increasing international travel.

Secondly, as we pointed out earlier, in the **close confines** of an airplane cabin, it is the immediate local environment at any time that really counts. You're up close and personal with your immediate neighbors and their space becomes your space too. What they put out into the air, you are likely to take into your lungs. These are rather close encounters, after all.

Thirdly, concerning disease transmission, all **surfaces** in an airplane can be contaminated with germs that can make you sick. According to the CDC, viruses can survive for short periods in the air and on surfaces like tray tables, seat backs and arm rests.(17) When you think of it, you are vulnerable to all the hygiene habits (or lack thereof!) of all the other passengers on board. That too, is another scary thought. Enough said.

Finally, it is clear that **the incubation time** of most airborne infections is much longer than the time spent on the aircraft. Illnesses would usually become clinically apparent only several days after air travel has been completed, and everyone has gone their own way. Therefore, it is reasonable to expect that the source of any possible infection and the association with air travel are not easily identified. Travelers (and their doctors, too) often assume that illnesses associated with travel or vacation are consequences of some exposure (real or imagined) at points of destination.

In light of all these considerations, you should probably not take comfort in the AMA position. But if you are still nonchalant about this issue, here are some more tid-bits that should be more cause for concern:

- Known infections transmitted as a result of air flights include: measles, small pox, colds, influenza and most recently, tuberculosis.(18)

- Risk is increased when the aircraft is on the ground and loaded with passengers. Necessary ventilation is often turned down, if not off.

- The longer the trip, the greater is your risk.

- In 1997, a major study by a TV station in Columbus, Ohio found that 78% of airplane surfaces were contaminated with germs that could cause disease.

- Several countries (in Latin America, the Caribbean, Australia and the South Pacific) require that an insecticide be sprayed on all arriving international flights in order to kill any hitchhiking alien insects that might damage local crops. You would have to seek in advance, written medical exemption to get off before spraying begins.

- High CO2 levels in the cabin during take-off and landing caused the FAA to impose a limit of 1,000 ppm for all new transport aircraft, since 1997.(19) This has not always been the norm, and in some cases, it probably is still not so. Above this level, CO_2 will make you drowsy at least.

Playing Defense

Given what you have just learned about airplane travel, what defensive actions might you take to minimize your risk from exposure to possible contaminants in that restricted environment? You cannot control the condition or behavior of your fellow passengers, nor can you control the type of filters the airline will use or the amount of 'fresh' air that is brought into the plane. There are no windows to open, if you think about it.

So, what can you do? On the following page, you will find ten simple defensive steps that you can take when flying.

TEN DEFENSIVE STEPS WHILE FLYING[20]

- **Avoid flying when you're not well.** That's for your own safety and the health of others who travel with you. If you must travel with a respiratory condition, wear a mask! No one will begrudge you for it.

- **De-plane whenever you can**. On stopovers or unforeseen delays, wherever you have the option, get-off the plane. Try to get some real 'fresh' air before the next flight.

- **Report any suspicious problems**. If you detect any unpleasant odor; if the air feels rather stuffy; if you're sitting around someone who appears sick; if you have any kind of health apprehension inform the crew and let them orchestrate defense.

- **Wash your hands and wash them often**. Try to minimize contact with tray tables, shutters, overhead bins, arm rests etc., and avoid touching your eyes, nose and mouth in-flight.

- **Do not wear any strong fragrances**. Be considerate of others who might have sensitivities.

- **Carry a small mask** and apply it in the worst case scenario, to protect yourself on board.

- **Drink lots of fluids** (pure water, preferably) but avoid alcohol and caffeine, both of which can be dehydrating.

- **Favor non-smoking international flights.**

- Remember to **take any prescribed medications** on board.

- **Utilize ionization technology** by wearing a personal air purifier as your 'buddy.'

The personal air purifier will be at work for you, disinfecting the air in your immediate surroundings and helping you to breathe the same high quality 'fresh' air that you must be getting accustomed to by now.

This is a proven defensive method against whatever you might have to contend with in the air, next time you fly anywhere. Researchers at the University of Cincinnati have studied the efficiency of the personal air purifiers in a laboratory setting that models quite well the situation of a seated passenger on board a commercial flight.(21) The data clearly suggest that the unipolar ionic air purifiers are particularly efficient in reducing aerosol exposure in the breathing zone when they are used inside confined spaces with a relatively high surface-to-volume ratio, as in typical aircraft seating areas (as well as inside automobiles, for that matter).

Are these products really worth it? Indeed, do they even work? Well, at least some travel experts think they can't hurt, if only to ease people's fears about traveling to new regions. With outbreaks of emerging diseases like SARS, the potential threat of Avian bird flu and the anticipated long-overdue pandemic in the wings, these fears are not unfounded. Overall, the number of Americans venturing overseas has gone up over 50% in ten years, sending travelers into regions where good medication and doctors aren't always available.

Actual user comments:

"Love it, my security blanket."

"It is great on airplanes."

"Allows me onto airplanes."

"Makes travel easier."

"Thank you!! I can fly again!!"

"I haven't been sick after airplane flights since using this."

"Excellent on airplanes."

Arnold Barnett, a professor from the Massachusetts Institute of Technology who has conducted studies about transportation safety, has been quoted in *The Wall Street Journal*, saying that ' as long as these gadgets don't make people obsess about their hang-ups, they're better than just sitting there and being scared.'(22)

But they do much more than comfort the distressed passengers and calm their anxious nerves. They deliver the benefits of advanced technology. Here's what one airplane cabin manager had to say about the value of the personal air purifier in her traveling experience.

"I used the purifier on the A320 aircraft only, as this is the aircraft we have a lot of problems on.

On many occasions where we had mild oil fumes through the cabin, I found the purifier worked well. Normally I would suffer headache, burning throat, dizziness and on some occasions - nausea. I did not experience any of these symptoms.

On one occasion, on descent we experienced oil fumes. My colleague sitting next to me had a headache and dry burning throat from the fumes. I gave her the purifier to use for 5 minutes. Her symptoms completely disappeared.

We also find after long flights on these particular aircraft we feel completely exhausted. Using the purifier I felt fine after these flights - no exhaustion whatsoever.

If one of my work colleagues had a head cold, I also used the purifier. I did not catch their cold. Not sure if this is my immune system or the purifier.

In summary the purifier certainly improves our air quality and we would like to see it used further in our industry."

J. Newlands, Cabin Manager

Constance Stallings, a 67 year-old copy editor blames germs circulating in the close quarters of planes for the respiratory infections she seemed to develop every time she flew. After trying dozens of nose ointments and moisturizers she stumbled upon a personal air purifier and put it to work for herself. Now she just ignores other gawking passengers while she lets it clean her breathing zone throughout every flight. "I put it on right after the security checkpoint so I won't forget", she says. "I haven't got sick once since I started using it."(22)

That's a useful anecdote and there are many, many more, but that's not the weight of the evidence. From the series of papers coming out of university research laboratories, and published in peer-reviewed journals, there is no doubt remaining that this method of protecting your personal space, especially when traveling, by car or by plane makes consummate sense.

It is a personal choice that you must make and a practical step that you can take to minimize your risk of airborne infection and chemical pollution in your traveling space. In a word, you should not leave the house without it and certainly not leave the ground without it.

You may be a frequent flyer or just the occasional airline passenger, but sooner or later, you've got to get back to work. So we move next to cleaning up your ***WORK*** *Space.*

FOURTEEN

Clean-Up Your WORK Space

"Air Pollution problems in the workplace demand a Spectrum of Answers."

You have now considered defending your *personal* space, maintaining a healthy *home* environment, and protecting your lungs as you *travel* by car or airplane. But next to your own home, you will spend most of your time at work. Americans in fact, spend about 28% of their time, on average, in their work environment.

Buildings should always provide an acceptable indoor environment to support the well-being and productivity of the work force. It is the responsibility of employers and building managers to maintain an acceptable level of indoor air quality. But what that level is depends strongly on who must accept it. At least four different stakeholders will have their different perspectives.

WHOSE PERSPECTIVE?

Employers will be generally satisfied if there is minimal absenteeism and no obvious loss of productivity on the part of workers because of the negative influences of inadequate air quality. Failing this, there could be substantial **economic costs** to employers. As we reported in Chapter 1, this aggregate value across America runs into the tens of billions of dollars each year.

By this criterion, employers ought not to be satisfied with the status quo and should be examining ways of investing in building renovations and maintenance to improve air quality. This can be cost-effectively managed so that the return-on-investment pays off in the short term as workers respond to the improved environment.

If you are an employer, it's presumably your responsibility to ensure a safe and healthy work environment. That involves providing good air quality that serves the needs of your employees. Failure to provide a safe and healthy work environment could expose you to legal remedies for any health consequences or associated damages that could be legitimately claimed.

Building owners and managers are inclined to be content as long as tenants are not making complaints, or demanding different requirements for ventilation, or temperature and humidity control. They also have an interest in making sure that their tenants are not engaging in different kinds of polluting activities. But when the tenants become restless and begin to assert their tenant rights and complain, that sets up a case for **legal action** and attendant costs, and possibly increased insurance costs. Building managers can do much to affect indoor air quality by aggressively pursuing a rigorous detection and maintenance program. Otherwise, they could be fighting with unhappy campers or else losing tenants, if not more.

Government regulatory agencies like the U.S. EPA, have no real jurisdiction indoors, but from a public health perspective, they must seek to minimize exposure to any kind of toxins in the work environment. The National Institute for Occupational Safety and Health (NIOSH), part of the Department of Health and Human Services, reviews scientific information, suggests exposure limitations, and recommends measures to protect workers' health. The Occupational Safety and Health Administration (OSHA), part of the Department of Labor, sets and enforces **workplace standards**. Altogether, these government agencies must work to minimize any adverse health effects in the work environment and strive to eliminate comfort complaints. The failure to enforce reasonable standards for indoor air quality in the workplace can only mean a dramatic impact on public health costs. It's simply a matter of *now, or later.*

Employees are the most important stakeholders in the quality of indoor air in their working environment. After all, it is their **health and well-being** that is at stake. It would not be surprising then that they tend to demand the most rigorous standards of all. Every employee would like to work in a clean, dry, well-ventilated environment that is thermally comfortable, with no unfamiliar or objectionable odors in the air, and free of any invisible toxic chemicals and fine particles. Anything less not only tends to make the physical conditions uncomfortable, but in real terms, it threatens to cause short-term illness and perhaps long-term suffering of employees. That means, most of all, the possibility of substantial personal injury and health costs. This, no employee should be willing to tolerate. As an employee, your health may be at risk from poor indoor air quality at work.

But having said all that, the typical working environment in America still leaves much to be desired. Workers still have to deal with exposure to unhealthy contaminants and uncomfortable conditions on the job. That makes a significant contribution to indoor air pollution.

INDOOR AIR POLLUTION ON THE JOB

When the 12,500 occupational and environmental health professionals who belong to the American Industrial Hygiene Association were asked in a survey about important health issues for workers, more than 40% named indoor air quality as the first or second most important. Again, when 1000 full-time workers were interviewed in July 2000, as many as 75% ranked the quality of air at work as very important.(1)

Pollution of the typical office air environment can result from poor or inadequate ventilation, lack of environmental control (temperature and humidity), elevated levels of pollutants and their persistent sources, different activities and processes within the building, as well as contaminated outdoor air coming in. The net result though is that industrial workers and office employees have to contend with unhealthy substances in their breathing zone and beyond.

The building you work in is subject to much of the same pollution problems that you face at home. Building materials, carpets, second-hand tobacco smoke, infectious microbes spread through respiratory aerosols, and other sources - all present the same challenges as at home. But in addition to those, some problems such as scents and fragrances, automobile exhaust (from loading docks or nearby parking lots), industrial chemicals and cleaning solvents, manufacturing activities etc. can be much more common in the workplace or office. The most important pollutants that affect the health of those in the office fall into the same three major classes that we have identified earlier: biological pollutants, chemical pollutants and fine particles.

U.S. companies could save $58 billion annually by preventing sick building syndrome.

Biological pollutants include mold and allergens, bacteria and viruses. Offices can be especially vulnerable to microbes, because fungi and bacteria find nourishment in inadequately maintained humidification and air-circulation systems, and in dirty washrooms. Contaminated and poorly functioning HVAC systems are notorious contributors in this regard. Of course, molds and allergens in office buildings present the same increased risk as in the home, for allergic reactions and respiratory disease.

The molds of concern include *Aspergillus, Penicillium, Stachybotrys* and *Fusarium. Stachybotrys* is a particularly dangerous type of mold which has increasingly been found at many workplaces in recent years. It likes to grow on high-cellulose, low-nitrogen materials like fiberboard, gypsum board and paper. So it is often found in libraries and rooms where papers are stored. This greenish-black mold releases spores which contain potent substances that are called mycotoxins. When released into the air, they can be quite dangerous. Many people are allergic to them. Some develop hay-fever-like symptoms while others who have chronic respiratory disorders can have trouble breathing and could even become disabled by them. As you recall now, the key to all types of mold control is always moisture control.

Allergens of concern result from the same notorious dust mites, cockroaches, animal pests and molds. These can also trigger or exacerbate asthma, which you recall affects more than some 17 million Americans.

Bacteria and viruses are commonly shared among office workers. The employee who stays home with the common cold or 'flu' and any of the other respiratory ailments can sometimes be the exception more than the rule. Many a person who comes down with a contagious illness can trace it back to the workplace where they were exposed, often inadvertently.

Chemical pollutants are many and varied in the work environment. Most buildings will have anywhere from 50 to 300 different volatile organic compounds in the air. Formaldehyde and the solvent toluene are probably the most prevalent of all these VOC's. The major sources of formaldehyde are likely to include particle board, fiberboard and plywood in furniture and paneling, carpeting and glues.

Often VOC's can be released from paints, adhesives, solvents, upholstery, draperies, spray cans, clothing, construction materials, cleaners, deodorizers, toners for copiers, felt tip markers and pens, and correction fluids. And that's not a complete list. Factory workers and others whose jobs necessitate the use of large amount of solvents and manufacturing chemicals are much more likely to risk exposure to VOC's. One way to help minimize exposure is to always put toxic operations like painting etc. in separate, closed-off areas with their own exhaust ventilation systems, and to have full protection for the workers involved.

Taken together, VOC's are quite common and their indoor levels may be anywhere from ten to thousands of times higher indoors than present in the outdoor air. Many of these have been found to be irritants to humans and contribute to the sick building syndrome (SBS) that was discussed back in Chapter 4. Generally, exposure is low but the consequences of cumulative exposure to different chemicals over many years could be more serious than we can practically quantify. We cannot exclude their contribution to many chronic and degenerative conditions.

Particles of dust and dirt can be brought indoors from the outside air and can be generated by activities such as smoking, remodeling, office machine operations, and aerosols. But even in the absence of all these, there is the normal shedding of skin that is a major contributor to dust in home and offices.

The small particles that you cannot see are more likely to be harmful to your health. As we have pointed out several times, it is especially the very fine particles that are most dangerous. They are inhaled into the lungs and contribute to lung and respiratory disease. In fact, as we did mention earlier in the book, fine particles are associated with daily mortality in America.

It is therefore a progressive social phenomenon, that a smoking ban in the office has become almost absolute in many industrialized communities, although no one knows what often goes on behind closed doors. If you ever have to contend with second-hand smoke, just remember that breathing second-hand smoke consistently, more than doubles the non-smoker's risk of lung cancer, and that's not all.

EMPLOYEE HEALTH AND COMFORT

In Chapter 4 we discussed several building-related illnesses (BRI) which have more specific cause and effect relationships. In contrast, we also looked briefly at the sick building syndrome (SBS) where occupants experience acute health and comfort effects that appear linked to time spent in a particular building, but no specific illness or cause can be unambiguously identified. We also mentioned the more recent understanding of the environmental illness phenomenon where some patients show extreme hypersensitivity to multiple chemicals in the environment (MCS).

The sick building syndrome is a real factor in the lives of many workers on their job. Many studies of this SBS phenomenon have been done over the last four decades and the results prove somewhat contradictory. But some 'risk factors' have been unambiguously established. There have been *consistent* findings of association between increased prevalence of symptoms and air conditioning, job stress/dissatisfaction and allergies/asthma. There have also been *mostly consistent* association of symptoms with low ventilation rate, carpets, occupant density, video display terminal use, and female gender. *No association* with altered symptom prevalence was found for total fungi exposure, viable bacteria, particles, air velocity, carbon monoxide, formaldehyde and noise.(2)

The classic California Healthy Building Study in the early '90s' examined 12 buildings with 880 occupants and found a wide variety of symptoms.(3) Eye, nose and throat irritation were the most frequent. Occupants of buildings with central HVAC systems were more symptomatic than those in buildings with natural ventilation. Workplace factors, like the presence of carpet, the use of photocopiers and carbonless paper, space sharing and the distance from windows, were associated with more symptoms. Later analysis by other researchers demonstrated for the first time, a link between exposure to low-level volatile organic compounds (VOC's) from specific indoor sources and symptoms of SBS.(4)

At about the same time, a Library of Congress Study examined four buildings in the Washington, DC area, where 6771 employees of the EPA and the Library of Congress worked.(5) The variables they identified to be most closely associated with health factors included:

- Comfort characteristics (dry air / dusty office and hot, stuffy air)
- Odor factors (paint/chemical and fabric odor)
- Individual susceptibility factor (dust/mold allergies and sensitivity to chemicals)

Many 'problems' or 'complaints' associated with building environments have been reported for years to the National Institute for Occupational Safety and Health (NIOSH). The institute has responded to thousands of requests for building investigation. However, the media, perhaps inadvertently, gave a boost to the public awareness of this problem when a national news program, back in 1993, discussed health effects associated with non-industrial indoor environments. The immediate response was overwhelming. NIOSH received about 500 requests for indoor environmental quality (IEQ) health hazard evaluations across the country. They selected a subset of these, investigated 160 offices, schools and other non-industrial work settings and analyzed some 2435 questionnaires.(6) They found some interesting associations:

- Multiple lower respiratory symptoms were positively associated with outdoor intakes located close to pollutant sources and with all measures of uncleanliness.
- Building maintenance showed a decrease in multiple lower respiratory symptoms, but in some cases, an increase in multiple atopic (allergic) symptoms and diagnosed asthma.

The U.S. EPA has recently developed a standardized sampling protocol to measure the indoor environmental quality in non-compliant office buildings. They have since created the Building Assessment Survey Evaluation

(BASE) study that collected data on 100 representative U.S. buildings between 1994-98.[7] This is the largest and most comprehensive IEQ field study ever undertaken in U.S. office buildings and continues to provide valuable IEQ baseline information. The data is particularly useful since it is done on 'radon' buildings without specific complaints *per se*.

TESTING AIR QUALITY AT WORK

Since environmental conditions on the job, including air quality, can affect your health, you could (and therefore, should) take responsibility to be proactive in cleaning up your working space. You might begin by trying to find out what is the true quality of the air you're breathing daily at work. If you have any reason at all to be concerned about the air quality in your particular working environment, then you should consider advocating for routine air quality testing to be done.

But only tests that are designed and validated for the indoor environment should be used. Most of the typical occupational, hygiene tests are not really applicable for indoor office environments. They lack both the sensitivity and selectivity needed to properly monitor indoor pollutants. So, be sure that only tests specific for indoor environmental study are used, and only by laboratories that have accreditations and this kind of proven experience. Such labs should be ISO registered and have the necessary accreditation for environmental mold analysis where necessary.

Actually, there are several test kits commercially available that allow building managers, safety and health officers, and even building occupants to easily evaluate their environments for primary air pollutants. You could even order one of these kits. They are generally easy to use, with explicit instructions and the kits would be sent back to a laboratory for complete analysis. Usually, you will receive a simple report back, describing the results and showing you how the air pollutants in your working environment compares to the recommended levels. These results may appear normal or they may indicate that a potential issue exists which needs to be more fully evaluated.

That might be a convenient first step for you to take. More extensive tests are available for the advanced consultants or scientists, and these would normally require the use of more specialized materials and equipment. They might cover tests for molds and allergens, bacteria, aldehydes, pesticides, VOC's, microbial VOC's, among others. Certain tests, known as environmental chamber tests, are also available for measuring gases that emit from products. These degassing tests are available wherever information on a particular product is needed, or if one needs to find out why a particular product in a building has an odor or is contributing to other air quality concerns.

If a significant air pollution issue is found to exist in your building or there is a known health concern, then you ought to insist on professionals coming to your workplace to conduct a full investigation. These professionals would be skilled in observing your building thoroughly, talking to affected workers, identifying potential sources and issues, evaluating the practical ventilation performance, conducting sampling and analysis, and finally, bringing some resolution to the particular air quality problem that you are obliged to work with.

Workplace

Health & Safety

demands

Purified Air ...

nothing less!

Testing unveils the writing on the wall! Low grade filters may be clogged and neglected, their usefulness having ended a long time ago. Maybe the ventilation system has been weighed in the balance and found waiting. Perhaps some building or furnishing materials are contaminated and should give way to others by replacement.

One of the most important factors in protecting yourself in any working environment is to keep constant watch on the operation of the central HVAC system in your office building or plant and report any malfunctioning or maintenance problems.(8)

When properly designed and operated the HVAC system should:

- Control the temperature and humidity in the building to provide a decent, comfortable human working environment.
- Distribute enough clean outdoor air to meet the ventilation needs of the people working inside.
- Remove odors and pollution from the air in the building.

However, during the past thirty years or so, many plants and office buildings have had troubles with their HVAC system, causing people to experience drowsiness, headaches, allergies, respiratory and skin conditions. You don't have to be an expert to decide if the system is doing its job, even though you may need a professional to get the system back in proper working order and to replace worn-out or defective equipment. Below are some questions to ask, and problems to look out for in your own workplace:(8)

Does your workplace have a ventilation system? Many workplaces have a poorly designed ventilation system. This is especially true if the plant or building you are working in was built for another purpose, and then converted or adapted to its current uses. That is not uncommon. If you are working in a large plant area, for example, it may have few or no walls or partitions. Look at the number and sizes of exhaust fans along the walls and in the ceiling, and look at the number and sizes of the windows and doors which are supposed to let air in? Are there enough of them? Are they properly placed so that everyone working on the shop floor can get enough fresh air?

If you are working in a closed office or workroom, look around the room for air vents. These are usually small grills on the walls or floorboards, or diffuser panels in the ceiling. There should be at least two vents in each room, one to supply air and one to remove it. If you don't have any vents, then the only air movement and air replacement in the room has to come from open windows and open doors. If you have only one vent, look for the other. The other one may be in the room next door with a small opening for air, say next to the window panel, connecting the two rooms. Or maybe there is only a partial barrier between offices and the air flows around or above the partition(s) from one office to the other. Or, of course, the other vent may be down the hall, and of no help for your room when the door is shut. Remember, air is a fluid which flows from one place to another - it can't just enter or leave from your room through one single vent and stop there. If it really has no place else to go, it will just sit as a stagnant pool, with no movement until the door or windows open. That's a basic, but often respected principle.

Are the room vents supplying or removing air? For the two vents in your room, or the two working together in connected, neighboring rooms, one (the supply vent) should be supplying air to the room(s) and the other (the exhaust vent) removing it. How can you tell which is which? Just put a piece of tissue paper up to the vent and see which way the air there is flowing. For the supply vent, it's often useful to just tape the tissue to it, so you can always just look up and see if air is being supplied. (Some air-supply systems in large buildings are timed to go off after working hours. Make sure they don't go off at 4 or 5 pm, when you or others are scheduled to work until 6 or when you have to work overtime or on Saturdays. If this is happening, speak to the building staff [or building health and safety person, if there is one], and if that doesn't bring results, bring the problem to the attention of your local health and safety committee or officer.)

Are the supply and exhaust vents adjacent to each other? If this is the case, the air flow can get "short-circuited," that is, the air can flow out of the supply vent and go immediately into the exhaust vent without adequately circulating in the room. If this is the case in your workroom, one of the vents needs to be moved. Usually it is easier to move the exhaust vent, since this vent often returns the air through the dead area in the false ceiling above the room. So to change its location, the building staff may only need to move the diffuser panel in the ceiling from one place to another. (The supply vent is usually fed by an air duct made of sheet-metal pipe, and so is more difficult to move.)

Are the vents blocked in any way? If file cabinets or bookcases or just piles of paper block any of the room vents, then of course they are going to block air flow. Also partial room barriers, designed to give you some privacy, block air flow. So check out your office or workroom, remove any blockage you can and ask for help through your health and safety committee to get file cabinets, etc., moved.

Are there dead spaces in your workroom or office where people regularly work? In dead spaces, air pollutants can accumulate and the air becomes stale and uncomfortable. Check for these by looking at your workplace and figuring out how air usually flows. If you feel or suspect that there is a dead area in the room, you can then get an inexpensive smoke tube, break off the tip in the suspected dead area, and see if the smoke drifts up (bad, no circulation) or moves out of the area (a positive sign of air movement). Also, if there are no flammable materials in the area, you can just light a match and see where the smoke flows.

Are the temperature and humidity levels adequate? Most government agencies and professional groups recommend that air temperatures in an office or workroom remain between 65 and 75 degrees Fahrenheit throughout the year. In plants with large open areas and in some offices, temperatures may drift below 65 degrees in the winter, and space heaters may need to be supplied. Obviously, they also need to be placed away from flammable materials, so that they do not become fire hazards.

The humidity in all areas should be reasonably low, to prevent mold buildup, but high enough to avoid dry noses, headaches and susceptibility to colds and flu. Guidelines vary, but most groups recommend a range of indoor humidity of between 20 and 50 percent, or perhaps 30 to 50 percent. It's useful to install a wall device which measures both temperature and humidity, and to use it as a basis for getting building staff to improve your indoor air quality. If you don't get satisfaction, record room temperatures and humidity at regular times during the day, and report these to your local health and safety committee.

A SPECTRUM OF ANSWERS

The same basic strategies apply for cleaning up your surrounding air at work as they do at home. The same space-age technology can be utilized to purify the air in your office or on the factory floor. The RCI™ technology will provide the same *spectrum* of answers, when it is used to remove biological contaminants and volatile organic compounds that pollute your working space.

In principle, you could actually begin your strategic defense against workplace exposure to airborne contaminants by protecting your immediate breathing zone with your **personal air purifier** worn about your neck while on the job. That will afford you all the benefits that we described and justified fully in Chapter 11.

If you work in a very small office, you might create your own **fresh air supply on the go**, by taking the same traveling unit that you use in your automobile as you commute to work and placing that on your desk, for example, to get the local benefit of that RCI™ space-age technology application in your office. That is a convenient and versatile utilization of the traveling air purification device.

But a third option and perhaps a better one, is to create a constant fresh air environment in your office and to keep it that way by placing a stand-alone unit that utilizes the space-age RCI™ technology to continuously provide **the spectrum** of air purifying solutions.

Silent, radiant air purification would mean that there's no noisy moving parts, no maintenance and no filters to clog up and replace. Radiant catalytic ionization will provide you state-of-the-art highly effective technology for reducing mold, bacteria, viruses, volatile organic chemicals, and in fact, the whole spectrum of air contaminants in your working spaces.

"It's a mere illusion that the solution to pollution is dilution!"

Imagine yourself in your office, in the healthy atmosphere and perhaps the stillness that follows a thunderstorm - a quiet world, refreshed and cleansed by naturally occurring ionization and oxidizers. The atmosphere is almost divine. Every breath you take would then be a pleasant reminder of what truly fresh air should be. It's the sensation of the natural outdoors brought inside. And all the while you work, you experience the same exhilaration of the air you breathe as though you were in the mountains, or standing by a waterfall or on the beach. It's one breath of fresh air at a time, even while you work. A complete spectrum of contaminants have been removed from your work space, leaving you that fresh air to enjoy!

If you work in a large office or on an open factory floor, then you may want to go further. You may need a larger, more powerful unit that can be placed near your work station but even provide benefits of purified **Fresh Air** to your coworkers all across the floor. And if that were not enough, you might go yet another step to introduce the space-age RCI™ technology directly into the ***Duct Work*** of your building's control HVAC system.

That's the subject of the final chapter, as we address how to **influence your COMMERCIAL space.**

FIFTEEN

Influence Your COMMERCIAL Space

*"To **'think globally but act locally'** makes as much sense indoors as it does outdoors."*

We have been considering the application of air purification technology to expanding spheres of human activity. We began with defending your PERSONAL space and examined the real value of a personal air purifier to provide an electrostatic shield to your breathing zone.

The second sphere was in your home where you would typically spend most of your time. There we explained how the space-age RCI™ technology could maintain a healthy atmosphere in your HOME space. You could provide up to 3000 square feet of living space with Fresh Air continually, for the benefit of your entire household. But you would not remain at home all the time. Therefore, as you venture out into a wider sphere, you could defend your TRAVELING space with your personal air purifier, particularly when flying, or experience fresh air as you go with a mobile unit that attaches to your automobile dashboard for the benefit of all who travel in your vehicle.

Once you get to work, you have several choices in that even larger sphere of activity. If you have only a small office or work station, you could rely on your personal air purifier about your neck, or the same mobile unit from your car. The latter could sit on your desk to provide more fresh air as you go about your daily schedule. A larger office might require a spectrum of answers to the challenges of air pollution. If you had to work on an open floor, then that could require a source of Fresh Air that might operate even 24/7 for maximum efficiency and health in your WORK space.

Beyond that, there remains your exposure to a wider sphere of activity as you go about your duties from time to time. That is your COMMERCIAL space which you could seek to influence in one way or another. This commercial space will at some time include healthcare facilities which you must visit sooner or later; schools that your children or grandchildren must attend; hotels or other hospitality venues when you must find accommodation away from home; entertainment facilities where you will find recreation and relaxation when you need either of them and other places of interest that you will attend as the need arises. Each of these commercial venues has its own particular challenges and demands for solutions to indoor air pollution problems.

We will address these various commercial spaces in turn, beginning with ordinary healthcare facilities.

HEALTH RISK IN THE AIR COMES WITH HEALTHCARE

Healthcare facilities, such as hospitals and doctors clinics, must pay particular care to indoor air concerns. Microbes in the air present a major challenge both to patients and visitors, as well as to the dedicated staff who are privileged to work there. All these groups are placed at risk of infectious disease just by virtue of being in a healthcare environment.

Some patients are obviously more at risk than others. Those with chronic disease, respiratory illness, immune compromise or suppression, the frail or elderly, the fragile babies, transplant patients, the post operative patients or those with open wounds - all these and perhaps more are highly susceptible to airborne infection.

Different parts of a healthcare facility may also show particular vulnerability. The emergency room is obvious. Patients coming in may not be aware of their infectious condition and even before triage, they may interact with others in the ER and unwittingly transfer infectious agents to others. Some wards may be characteristically contagious such as infectious disease units or gastrointestinal areas. Other parts of a healthcare facility may have to deal with volatile organic compounds just by virtue of the materials in common use.

The fact is that despite the best efforts, about five percent of all patients (1 in 20) who go to hospitals for treatment will develop (what you might remember from Chapter 4, is called) a 'nosocomial' infection while they are there. That is the result even though hospitals have infection control committees, staff who detail and supervise special protocols and precautions to help prevent infections and minimize their spread in each facility. The fact is that microorganisms are 'hell-bent' on survival just as we are and they will do whatever they have to do. In the process we become the victims of their ingenuity and desperation. Therefore, the general idea is to stay out of hospital *for health's sake*, as long as you possibly can (and your doctor wills).

Before we even get to hospitals, take first the typical doctor's office. Everyday the staff must contend with whatever contagions the patients bring to that office. With little information to screen each patient a *priori*, everyone in the doctor's office is at risk. People are constantly coughing and handling stuff in the office. Children run around and play as if all were well with their little world, oblivious to all the possible biological hazards that surround them and their parents. Seniors wait in line to be attended to while they must survive whatever is in that enclosed air space.

Everyone in the doctor's office is at risk.

There is no escaping this potential infection reservoir unless proactive steps are taken. First, each staff member (doctor, nurse or support staff) might do the sensible thing and take precaution with a **personal air purifier** to help protect their breathing zone, since they must live with the potential threat on a daily basis. Secondly, the RCI™ space-age technology could be taken advantage of, to provide **Fresh Air for Life** and help protect all who enter the clinic. Thirdly, the ventilation system (central HVAC) must be scrutinized and maintained, so that adequate ventilation is provided, and humidity is controlled to minimize growth of bacteria and mold.

The doctor's office is bad enough, but then there is the notorious **hospital environment.** That is a necessary *health* care facility - one that cares for the sick yes, but by its very nature, it is also a reluctant source of *infection* by itself.

Begin as you enter an emergency room. All that was true in the doctor's clinic is now exaggerated here. The patients tend to be more ill, more infected and infectious, more vulnerable to infection and often in more cramped quarters. The staff may well do themselves a big favor. Each doctor, nurse and support worker, as a means of self defense, would be wise to invest in a **personal air purifier**. Beyond that, space-age RCI™ technology could be implemented to advantage, bringing **Fresh Air for Life** to the benefit of all who must visit and wait there. Ventilation should be enhanced to minimize the mixing of room air and to increase displacement - sweeping the air out as much as possible and through a HEPA filter in the return duct to remove as many particles and bioaerosols as possible. For isolation rooms where infectious suspects are isolated following triage, the air should be exhausted 100% to the outside so there is no chance of the polluted (infected) air returning to the hospital.

Apparently, in Europe, there is a trend toward more natural ventilation. That is, they seek to use more open windows wherever possible. In the U.S., those hospitals that are mechanically ventilated must be supplied with 100%

outside air. How much air (and how clean that air is) will vary from one hospital to the next. Different parts of the hospital have specific ventilation requirements, depending on suspect microorganisms, hazardous chemicals and even radioactive substances.

Patients in hospital who are known to be infectious must be isolated and their rooms need to be well ventilated to the outside. Most hospitals have designed rooms with a *negative* pressure for this purpose, so that no air goes out into the hospital corridors or elsewhere to carry airborne infectious agents any further.

On the contrary, high-risk patients, like those whose immune systems are compromised, will have special rooms that are under *positive* pressure. Fresh air flows into the room from the central pre-conditioned supply and the buildup of the positive pressure ensures that no contaminants can come in from the outside by any other routes. These patients therefore become less susceptible to whatever airborne contagions may be in circulation inside the hospital.

Preventing

nosocomial (new) infections

may be

the biggest challenge

in any hospital.

The operating room(O.R.) is a special case. It is generally kept under positive pressure, again to protect the patients who are exposed and vulnerable in that delicate setting. Extreme infection control measures are adopted especially in the sterile field surrounding surgery or other invasive procedures.

In any case, the typical O.R. in America is required to be well ventilated - providing 30 cfm (cubic feet per minute) per person of fresh, outside air (according to the ASHRAE requirements for healthcare facilities). Most O.R.'s also utilize HEPA filters and often UVGI (ultra violet germicidal irradiation) lights for further air purification.

The surgical mask is designed to protect the patient from the surgeons and O.R. staff. But would it not make sense to enhance the value of that mask and the protection of everyone in the O.R., including the patient on the table, by using the space-age RCI technology to provide **Fresh Air for Life** routinely in the operating room? I would think so.

Infection is not the only airborne problem in hospitals and nursing facilities. But you would do yourself good, by wearing a **personal air purifier** to protect your own breathing zone if and when you had to visit inside any healthcare facility.

In hospitals and nursing facilities, it is also important to consider the necessary **chemical hazards** that produce volatile organic compounds (VOC's) in the environment. Routinely used cleaning products and disinfectants are a common source. Since most patients sleep at night, cleaning is often done during the day so more people are generally exposed. Special solvents are of interest, stemming from use in laboratory tests. These may include acetone, alcohols, benzene, formaldehyde, xylenes, methylene chloride and toluene. Some medical equipment may necessitate special cleaning with disinfecting agents like glutaraldehyde and ethylene oxide, to which some people are especially 'sensitive'. Anesthetic gases like nitrous oxide can accidentally get into the circulating air, and some volatile drugs (aerosols) like pentamidine (for AIDS patients with pneumonia) and some chemotherapeutic agents could also become a danger.

Then there is the cornstarch powder from **latex gloves**. These gloves became popular in the 1980's with increasing HIV-AIDS and other blood borne diseases. The powder is used inside the glove to ease putting on and taking off, and outside to prevent them sticking together in the box. As the gloves are pulled out of the box, the latex powder dust becomes airborne and can be inhaled. The latex protein molecules bind with the cornstarch powder to cause allergies which can at times be quite severe. Some 6-10% of the population have such allergies. Lately, there has been a distinct shift away from latex gloves but they are still quite popular. Non-latex gloves are provided as an alternative choice in health care facilities today. But if anyone on staff is using latex gloves, then others are at risk of allergy. Strategies of air filtration or ionization should then be appropriate. Again, for those very allergic to this, **a personal air purifier** would certainly be in order to protect their breathing zone.

YOUR CHILD DESERVES FRESH AIR AT SCHOOL

Hospitals are one thing, schools are another. On the subject of indoor air quality, modern schools in North America today do pose a challenge for school board authorities, teachers, children and their parents. Schools are unique because they are densely populated, so poor air quality can affect a large number of children. In addition, children are disproportionately impacted by these risks because they breathe more air per pound of body weight and have smaller lungs.

Using a 1999 U.S. Department of Education survey of 900 schools and other data, the EPA estimated that 30-40% of the nation's schools have indoor air pollution problems with contaminants that include mold, allergens, viruses, chemicals such as pesticides, and particulate matter.[1]

Another EPA investigation of 29 schools across the country found inadequate ventilation in most of the schools. Furthermore, nearly one in five schools (and by the way, one in fifteen homes nationwide) had at least one room with radon levels above the EPA's recommended action level.(2)

A list of possible hazards that threaten classrooms in many US schools could also include an abundance of mold, ceilings lined with cellulose tiles which make it worse, dampness and humidity from leaking roofs, condensation from low temperatures due to energy conservation and restricted circulation above the stained ceiling tile. In some older schools, you may find asbestos-containing floor tiles and dirty, worn carpets with all that that implies.

Exposure to indoor contaminants has exacerbated the asthma epidemic among children. The incidence has doubled to nearly one in ten in just two decades. There may be many causes but indoor pollutants are known to trigger asthma and the coincidence of 'tight building syndrome' after the energy crisis in the seventies points to a common factor: poor indoor air quality.

A former U.S. Department of Education facilities official was quoted as saying, "the classroom environment (in America) is a dirty secret". Many of the nation's school systems have been facing a budget squeeze, so they aim to save money by cutting down on maintenance. But reduced maintenance can create dust buildup, mold outbreaks, ventilation problems from dirty air filters - and subsequent health complaints. The *American School and University* magazine reports that spending on maintenance and operations as a percentage of school districts' expenditures has decreased continuously for several years. That simply puts more children, teachers and other support staff at risk.

An *Associated Press* report from Ohio (2003) began with this alarming observation: "For school officials in many districts, indoor air quality boils down to one simple issue: If you can't breathe, you can't learn".(3) Perhaps that's partly the reason why Minnesota requires its school districts to have indoor air plans and indoor air coordinators. Schools must evaluate building systems for air quality or jeopardize state funding. Each public school built in Minneapolis since 1997 has been designed to minimize chemical odors and maintain good ventilation. The result is fewer complaints about air pollution.(4)

In response to the national problem, the U.S. EPA has provided a voluntary Indoor Air Quality **Tools for Schools Kit and Program** (available at www.epa.gov/iaq/schools) since 1995. It is a program designed to assist school faculty and maintenance personnel in preventing and repairing indoor air quality problems. The kit is a fairly simple, common sense guide to recognizing some no-cost and low-cost solutions to reducing indoor air pollution from classroom animals, plants, pesticides, cleaners, art supplies and science materials.

More than 10,000 schools nationwide have adopted this program. It is both practical and inexpensive, and yields results. One independent school district in Galveston, Texas, for example, partnered with the University of Texas Medical Branch and got a 90% response rate. Of 54 surveys received at the outset, 42 classrooms had ventilation problems - 19 of which needed to have modified air conditioning maintenance to increase the flow of fresh air and to treat the air to ensure acceptable relative humidity. Additionally, 8 classrooms had evidence of leaks, 39 classrooms had cleaning problems, 28 had pest control problems, while 11 had exhaust fan problems.(4) That was just a start.

Fortunately, it was relatively easy and inexpensive to improve many of these problems by improving ventilation, fixing water leaks, forbidding tobacco smoking and choosing safer products. To prevent mold (one of the most common problems in schools) it is important to clean up water fast and remove all water-damaged items (including carpets) within 24 hours of a flood or leak. Where mold already exists, the affected areas need to be washed with soap and water, followed with a solution of 1 part bleach to 10 parts water, followed by thorough drying.

*If a child can't **breathe**,, properly, that child can't **learn** effectively.*

Other maintenance issues might involve regular filter changes, antibacterial tablets or UVGI in AC units, replacing carpets with tile and vinyl flooring, installing or replacing exhaust fans where necessary, perhaps masonry repairs and roof maintenance to inhibit water intrusion, and more baits and traps instead of chemical pesticides. These all lead to a healthier air environment for the teachers and children who have to spend some 30 or more hours in the classroom each week.

Daycare centers face similar as well as more acute problems. Smaller children in these environments have to constantly battle against airborne viruses and much more. They are particularly vulnerable to upper respiratory infections. A study of 1000 children in the U.S. showed that those who were enrolled in daycare before the age of six months had more respiratory infections and wheezing, than those who began daycare later.(5)

These children contend with more than infection. Airway disease can be caused by mold and other allergens. In Taiwan, for example, high temperatures and high humidity were shown to both favor these allergens. Workers in their daycare centers were more likely to suffer from SBS due to *Aspergillus*, with symptoms typified by nasal congestion, coughing, phlegm and fatigue.(6)

Children in daycare centers are more likely to have night time coughing and nasal congestion or rhinorhea (runny nose) than those in home care. They are more at risk of getting colds and earaches and of developing asthma.(7)

So what are you to do? If you have children in school or daycare, or if you teach in one of these contaminated environments, what precaution can you take?

You could become an activist like some parents and teachers have done. You could try to influence your school board or other authority to give higher priority to indoor air quality. You might insist on professional inspection and certainly on improved maintenance. Perhaps you can download the EPA's IAQ *Tools for Schools Kit* off the internet and go to your local principal, armed with it in hand to make some provocative suggestions. You could write or call your local political representative's office. You could collaborate with other teachers or parents who share your concerns, especially if they or their children have health symptoms that vaguely correlate with potential indoor air consequences. You might even enlist your family doctor's support and get him or her to write a short report with recommendations for addressing IAQ problems at school.

All that is good. But there is still a better way. As many environmentalists like to say: '*Think globally, act locally.*'

What that means is to think about the air quality in schools in your district, yes, but act specifically in the classroom where you must teach on a regular basis, or the classroom where your child must sit and learn for hours every school day. That's where you may have some immediate leverage and effect change.

If you are a teacher, you could (and should) get your own **personal air purifier** and wear it throughout your school day. That will be a protective shield as we have pointed out already in Chapter 11. You can thus effectively defend your own breathing zone. But you will do more than that. Whether you intend to or not, you will make a loud statement of protest throughout your school. You will get the attention of teachers, students and their parents, without saying too much or provoking political antagonism. But your example will speak for itself.

As a teacher, you may have the prerogative to introduce the same space-age RCI™ technology into your classroom for the benefit of all your students, as well as yourself. You can provide a *spectrum* of answers for all the various contaminants in your small classroom. If you teach in a larger open area, then you might need to introduce **Fresh Air for Life** for all to breathe in comfort and promote better health.

If you are a parent, then for smaller children the personal air purifier may not be a practical option. But for teenagers, that is a real possibility. With your child at any age, from daycare right through high school, it would be worth your while, knowing what you do now, to introduce the benefits of space-age technology to your child's teacher and enlist his or her support for providing **Fresh Air for Life** in YOUR child's classroom. You might want to partner with another parent or two and make this a joint project for the class together. You can lead a class project on one of those PTA days or simply make a class donation for the benefit of all.

There are faint signs of hope in the schools across America as administrators become more aware of the challenge of indoor air quality and the benefits to be derived from improving the physical learning environment. The school system of Greeneville, Tennessee for example, has installed RCI™ Technology in four pilot schools to date. They covered every classroom, hallway, restroom and common space within the schools to ensure 100% coverage of fresh air purification technology. The results were so impressive that a commitment has been made to provide this space-age technology to all the 17 schools that now constitute the Greeneville and Greene County School Systems of East Tennessee. Hopefully, that's a trend-setting example.

The bottom line is this. If you can't influence or change the entire school or school board, you might be able to change the air your child must breathe in school, in his or her classroom - today and every school day. As your child, he or she deserves to enjoy **Fresh Air for Life** at school.

THERE'S ROOM IN THE INN, BUT IS THERE FRESH AIR?

In today's fast paced world, everyone is on the move. Traveling from place to place, city to city, by road or rail, by sea and air - we're all going somewhere. Away from home, there is a need for accommodation. So hotels, motels, inns and the like have sprung up everywhere. Hospitality venues comprise almost 15% of all non-residential buildings in the U.S., according to the 1999 Commercial Buildings Energy Consumption survey.[8] We need a place to stay, to eat and to buy food.

But a quote from the October 1999 issue of *Lodging Hospitality* magazine tells an interesting story:

'The basic function of an innkeeper is to make his or her guests comfortable. That task includes clean, comfortable, quiet guestrooms; prompt and friendly service, and where applicable, fresh and tasty food. The innkeeper's obligation also extends to the air guests breathe. It should be as clean and free of objectionable odors as possible. Unfortunately, this crucial function is sometimes overlooked in designing, building and operating lodging properties.'[9]

That is so true. Every frequent traveler knows the experience of hotel and motel rooms where the air is foul not fresh, contaminated not clean, and often reminiscent of someone gone before. An article in *The Wall Street Journal* (1999) reported that 'the air quality in most hotel guest rooms is very likely more polluted and dirtier than most homes.'[10] In just an overnight stay, you could be exposed to a wide range of pollutants. Even if you request a non-smoking room, it is not uncommon to be obliged to stay in a room with the pervasive odor of cigarette smoke on drapes, furniture, carpets ... that noxious smell is residual in the air. And that's not the only nuisance you may have to put up with. Some contaminants may be far less obvious but no less significant. There are the VOC's from cleaning products and personal care products left behind, pesticides, degassing from carpets and fabrics, dust and dust mites, mold and other allergens - all of which can have consequences to health, either in an acute or cumulative sense. The presence of mold tends to be a particularly common challenge in hotel and motel rooms, with all their plumbing, frequent baths or showers and moisture problems. The typical symptoms that travelers experience - like sore throats, headaches and burning eyes - are likely reactions to poor indoor air quality, among other things.

Hotel design, construction and management must include indoor air quality as a critical variable from the very onset. So many factors contribute to this important environmental concern in this type of living facility. All the various elements need to be coordinated.

Fortunately, there are signs of a growing trend to more sustainable and green buildings, including hotels and motels. These service businesses recognize that their customers or clients are the only future they've got. With increasing competition too, they are obliged to address customer needs and complaints. As we have already pointed out, over 20 million Americans are reported as sufferers of asthma, and as many as 70 million with allergies. The number of people with multiple chemical sensitivities should probably run into the millions as well. Therefore, industry leaders must make efforts to make these millions of possible guests who have asthma, allergies or M.C.S. more comfortable.

There are some good pacesetting examples.

The Hilton O'Hare undertook a pilot project recently offering hotel rooms that address the air quality specifically, for guests afflicted with sensitivities to mold, chemicals and dusts.[11] They engaged an environmental company with this specialty to redesign several rooms. They gutted everything completely - from floor and wall coverings, to furniture and drapes, then to bedding and literally everything else. Then they rebuilt the rooms with special flooring, wall covering, fabrics, furniture, paints, adhesives and cleaning products - all designed to minimize VOC's, bioaerosols (mainly spores) and particulates (largely dust). They then installed RCI™ air purification technology in each room.

They even went further. They put in each room a unique monitoring system with an innovative, state-of-the-art approach to environmental monitoring. This provided the hotel with critical real-time data to help ensure healthy indoor air quality. The system recorded five crucial indoor air parameters: temperature, relative humidity, carbon dioxide, odor and gases (VOC's), and carbon monoxide. Constant monitoring enabled hotel staff to respond immediately, 24 hours a day, if any alert conditions would occur. Then, after each guest departed, each room was sanitized and purified using only products and amenities that were toxin free.

That's true hospitality service, if ever there was!

The Sheraton Rittenhouse Hotel in Philadelphia has labeled itself the nation's first 'environmentally smart' hotel.[12] They replace 70 percent of the air in each room every half hour. They used non toxic building materials wherever possible, including low VOC-content paints, adhesives and furniture. They also used recycled granite, nightstands made from old shipping pallets and wall covering made from fast-growing, sustainable bamboo. They deserve credit at lest for highlighting the trend, even if not starting a revolution.

Even in Hotels,

good Air Quality

is not a luxury ...

*it is a **necessity!***

One more example will suffice. The Natural Place Environmental Residence and Hotel in Florida was designed specifically for people who are chemically sensitive.[13] They provide rooms free of all known environmental allergens including oil-based paints, fragrances, antiviral fabrics and pesticides. They use no pesticides on the shrubbery, and guests are asked to avoid perfumes or cologne. Obviously, somebody was alert to the need and they are to be given credit also for their care and sensitivity in a marketplace where pragmatism and cost efficiency numb the senses of owners and managers.

As exemplary as these isolated examples may be, they are exceptional and rare. The typical hotel and motel in America today is still operated by managers unaware and insensitive to the challenges of indoor air pollution. You would be unwise to depend on the typical hotel or its management to look after your indoor air quality needs. Having read this book, you now know too much to be so easily deceived or ignored.

If you would avoid indoor contamination when you must find accommodation away from home, you had better be prepared to create your own fresh air as you go. **A personal air purifier** or a mobile unit, such as you use in your automobile, would be ideal for purifying the air in the room where you will reside. They are both practical options, convenient for traveling and will be effective in your *temporary home* space, as long as you need them.

ENJOY THE SHOW, ENDURE THE AIR

When North Americans are not at home, at work or traveling somewhere, we tend to congregate in places of entertainment and amusement. Perhaps it is a sign of affluence in society that we spend so much discretionary income trying to find some other person(s) to make our day - to fill our life with excitement and enjoyment and give us something to cheer about.

It's no wonder that many billions of dollars are spent each year indulging ...

- the producers and stars of Hollywood who stretch their creativity and skills to produce one big box-office hit after another, with Academy Award films that millions of cinema-goers love to see;
- the professional athletes in all four major American sports - Major League Baseball, the NFL, the NBA and the NHL - when they engage in pretentious warfare at our expense;
- the musical performances and rock concerts of every stripe and variety: from classical, religious or country music, to rap and rock, or hip hop and heavy metal sounds.

Add to that the racing and wrestling fans, the proud madness of college sports in secondary markets across the continent - and there's something for everyone. There's always some venue of entertainment appealing to our sense of loneliness, boredom and disenfranchisement, or the compelling social need to belong. The attendance records clearly demonstrate that we enjoy a good show.

In this cultural milieu, there is one common denominator that is usually taken for granted. Every spectator and participant in each of these entertainment facilities must, of necessity, breathe the air available in the stadium, theatre or rink, as the case may be. For better or worse, we must all share the same common indoor air environment. Unfortunately, in such establishments, the air quality leaves something to be desired.

No question about it, for a long time the primary indoor air contaminant in many entertainment facilities was second-hand smoke (ETS). As we saw back in Chapter 2, this was not without serious health consequences. So much so, that many state and local governments enacted legislation to ban smoking in public places such as sporting venues, (both indoor and outdoor), hotels and restaurants, bars and even casinos. Non-smoking sections did not do the trick, since tobacco smoke tends to migrate easily across all open spaces. Fortunately, smoking in public nowadays has by and large become an unsociable habit almost across the board. It's hard to believe that only a generation ago, smoking was 'cool', 'slims and thins' were in and 'light' was just right. Public health advocates have clearly gained much ground. But there are still some die-hards who flaunt and over-step their personal rights with total disdain and disregard for the rights of others. Such polluters ought to be avoided and reported.

In addition to ETS, there are at least two other classes of pollutants to be generally considered in entertainment facilities. The first is the occurrence of low levels of noxious gases that at times can permeate the arena, especially in indoor ice rinks for skating or playing hockey. Carbon monoxide and nitrogen dioxide are of special interest. These gases are products of incomplete combustion and can arise wherever there is incomplete combustion, engine malfunction or inadequate ventilation inside any recreation facility. Resurfacing equipment for ice skating rinks, certain types of motor vehicle events and even go-kart racing tracks - all tend to increase the risk of this type of indoor air contamination. NIOSH has recommended that average airborne levels of carbon monoxide should not exceed 35 ppm over a 10-hour period. Likewise, the WHO guidelines for nitrogen dioxide in the indoor air is limited to just 0.213 ppm as a 1 hour average. These pollution levels must be carefully considered by recreational facility designers and engineers, as well as event planners. The fans deserve no less.

In many entertainment facilities, the air quality often leaves much to be desired.

The other consideration of indoor air quality in entertainment facilities is that of infectious microorganisms. This is much more personally manageable. Amid all the shouting and screaming that goes on, is the generous sharing of respiratory aerosols that unsuspecting spectators must eject into the surrounding air. Who knows what viruses and bacteria are being brought into these venues for all the neighbors to inhale and take home as a souvenir of the game, movie or concert. It's a veritable 'free for all' and 'open season' experience. For those more vulnerable patrons like asthmatics, children and the elderly or the immuno-compromised, this poses an increased danger. Often, because of the induction period, it is difficult to attribute the infectious cause when one succumbs to the common cold, influenza or even more serious respiratory disease, including pneumonia.

The prudent defense is to use a **personal air purifier** to protect your immediate breathing zone in these 'free for all' meeting places where the risk of indoor contaminant exposure is increased. You need not be paranoid to take precaution where it's necessary and possible. Any risk-benefit analysis would confirm that your discretionary *wisdom in action* would be justified.

After all is said and done, you deserve to win your unseen battle against indoor air pollution in every sphere of activity that you enjoy.

Amen!

Epilogue

The secret is now out! For decades, a growing mountain of evidence regarding the importance of indoor air pollution has accumulated inside the esoteric libraries and laboratories of professionals. They have raised important concerns but primarily among themselves. Clearly, for any matter of such personal and public import, that was not good enough, by any standard. Therefore, we have sifted through some of the more significant and practical aspects of this research data and distilled it into the common marketplace of ideas, and that we hope is for the ordinary layperson's benefit.

Based on the knowledge you have gained from this book, you can no longer ignore the concerns raised. Now you understand that indoor air is far more polluted than outdoor air and like everybody else, you will spend about 90% of your time indoors - perhaps at home or work, and in your car or some commercial venue. You will only breathe whatever air there is available, for better or worse, if you do nothing to effect change.

True, the plethora of contaminants that pervade the indoor environment is still mainly invisible to the naked eye, but they are no less real. Surely, you now appreciate that **it's not just what you can see that threatens you, but also what you cannot see.** You're now sensitized to the reality that there's likely dangerous pollution in the common air and doing nothing is no longer a wise or credible option.

Failure to act on what you now know, will put you at risk for consequences that you can ill afford. Air pollution is more than media hype and activists' fodder. It is a real hazard that poses a present danger to your optimal health. Therefore, prudence dictates that you adopt at least some of the strategies outlined in this book.

You can *reduce* your exposure risk by taking proactive measures to minimize all the apparent sources of indoor air contamination in your immediate environment. You can improve *ventilation* at home by the appropriate use of fans, by strategic opening of doors and windows and by good HVAC maintenance. You can use the best HEPA *filters* available to remove as many particles as possible from your living spaces. That would be a good conventional start.

However, having done that, it's time to take advantage of the space-age technology available today. Protect your personal breathing zone by the judicious use of a personal air purifier that has been proven to form an effective electrostatic shield, keeping any polluted bioaerosols or fine particles away.

Use the RCI technology that exploits the best of nature's purifying methods and meets the requirements for air quality in space. Surely you appreciate the fact that it also dispenses a "purifying plasma" **to take the solution to the pollution** throughout your home or office. Also, be smart enough to defend yourself against the hidden threat that faces other automobile drivers and do no less to avoid the other threat when you fly commercially.

But remember you're not alone. Think first of those you love, who also need to become aware of their unseen battle against indoor air pollution. Protect your family for sure and those who live and work around you. Give them the opportunity also to enjoy the same fresh air for life.

Then become an advocate for indoor air quality in your sphere of influence. After all this, here are three worthwhile causes that you can champion as you seek to make the world a better place and your immediate environment a healthier one:

1. Lobby for change. Make your local legislators and policy makers more aware of the challenges that you and others have to face indoors as you seek out Fresh Air for Life. Push for legislation to monitor and control this closed environment, as much as we continue to do for the outdoors.
2. Defend your right. Work through your organized union, local residents' association, PTA, co-workers or whatever group has leverage to demand acceptable indoor air quality wherever you live or work. It is your right and you deserve no less.
3. Spread the word. Share this book with others. Wear your personal air purifier as a buddy wherever you go to also trigger conversation. It's a cause worthy of your focus and passion, and a mission deserving your time and attention.

In this overwhelming information age, think through your priorities 'Think globally, but act locally.' As you seek to maintain the best of health - ***one breath at a time*** - and as you strive to make a difference - ***one indoor space at a time*** - use the information gained from this book to **act in wisdom and treasure the indispensable gift ... Fresh Air for Life!**

REFERENCES

Chapter 1, Introduction

1. U.S. EPA, "The Inside Story: A Guide to IAQ," EPA Document #402-K-93-007, April 1995. Office of Radiation and Indoor Air (6604J); *see* also: U.S. EPA, "Targeting indoor air pollution: EPA's approach and progress." (1992) www.epa.gov.
2. U.S. EPA, "Analysis of the National Human Activity Pattern Survey (NHAPS) Respondents from a stand point of Exposure Assessment," Report 600/R-96/074. Washington, DC Office of Research and Development.
3. Lamacchia D, "Tightening Buildings to save energy can lock pollutants inside." Lawrence Berkeley Laboratory, Research Review, Summer 1991.
4. Marilyn Chase, *The Wall Street Journal*, Dec. 1998.
5. U.S. EPA, Report to Congress on Indoor Air Quality, 1989; *see also* U.S. EPA, Indoor Air Facts No.1: EPA and IAQ (Dec. 1991).
6. Abu-Shalback L. "The Impact of IAQ", *Appliance* (2000), 57(6): 53.
7. Winter J. "House, sick home & Indoor Air Pollution is definitely something to sneeze at." *Rocky Mountain News*, July 11, 2000: p3D (Denver Publ. Co.)
8. Quoted by Miller A, "Bad air breeds ailments in homes, schools, offices," *The Atlanta Journal-Constitution*, Sunday, July 20,2003.
9. Chelsea Grp. (7/31/00). "People are willing to spend money to improve IAQ." www.safetyonline.com.
10. Wallace LA, U.S. EPA. "Assessing Human Exposure to Volatile Organic Compounds." Ch. 33 in IAQ Handbook, Edited by: Spengler JD, Samet JM, McCarthy JF (2001), McGraw-Hill.
11. Fish WJ and Rosenfeld AH. "Potential Nationwide Improvements in Productivity and Health from better Indoor Environments," *Proc. Of ACEEE Summer Study* (1998), 8:85-97 (Asilemar).
12. Conlin M and Carey J. "The Business Case for Better IAQ," *Business Week* (2000), 5:114.
13. WHO, "IAQ: Organic Pollutants," Report on a WHO meeting, Euro Reports and Studies, 111 (1989).
14. Taylor ST, NIOSH. "Deluged by Sick Building Complaints," *Indoor Air Review*, Oct. 1993.
15. 59 Fed. Reg. 15, 968, 16, 006 (April 5, 1994).
16. See www.battelle.org.
17. *Battelle News Release*, June 6, 2001.

Chapter 2, Concerns You Can't Ignore

1. American Lung Association, *State of the Air 2005* Report.
2. Brook RD et al. "Air Pollution and Cardiovascular Disease," *Circulation* (2004), 109: 2655-2671. An AHA Scientific Statement.
3. "People with diabetes more sensitive to cardiovascular effects from air pollution." Natl. Inst. of Envir. Health Sciences (NIEHS). Press Release, May 2005.
4. Wallace LA, Nelson CJ and Dunteman G. "Workplace Characteristics associated with Health and Comfort Concerns in Three Office Buildings in Wash., DC." *IAQ '91 - Healthy Buildings*. Atlanta: ASHRAE; see also Wallace LA et al. "Perception of IAQ among Government Employees in Wash. DC,' *Technology: Journal of the Franklin Institute* (1995), 332A: 183-198.
5. Yu C and Crump D. "A review of the emission of VOC's from polymeric materials used in buildings." *Building and Environment* (1998), 33(6): 357-374.
6. Phillips M et al. "VOC's in breath as markers of lung cancer: a cross-sectional study," *Lancet* (1999), 353 (9168): 1930-1933.
7. Dockery TW et al. "An association between air pollution and mortality in six U.S. cities," *N. Engl. J. Med.* (1993), 329: 1753-1759.
8. Pope CA III et al. "Lung Cancer, Cardiopulmonary Mortality, and Long-term Exposure to Fine Particulate Air Pollution." J.A.M.A. (2002), 287:1132-1141.

9. Wallace LA. "The TEAM Study: Summary and Analysis." EPA 600/6-87/002a. NTIS PB 88-100060. Wash., DC. U.S. EPA.
10. A WHO-USAID Consultation. "Addressing the links between Indoor Air Pollution, Household Energy and Human Health." Washington, DC. May 2000.
11. World Bank Group. "Development and the Environment." *World Bank Report* (1992) New York, Elsevier NY, Publishers.
12. Quoted by Pat Veretto. www.frugalliving.com 07/05

Chapter 3, Ten Consequences You can't Avoid

1. Quoted by Miller A. "Bad air breeds ailments in homes, schools, offices," *The Atlanta Journal Constitution*, Sunday, July 20, 2003.
2. U.S. DHHS. "The Health Consequences of involuntary smoking." DHHS Publication (CDC) 1986, 87:8398. Washington, DC. US Govt. Printing Office.
3. California Environmental Protection Agency (Cal EPA), Office of Environmental Health Hazard Assessment. "Health Effects of Exposure to Environmental Tobacco Smoke." Cal. EPA (1997)
4. Jenkins MA et al. "Validation of questionnaire and bronchial hyper-responsiveness against respiratory physical assessment in the diagnosis of asthma," *Int. J. Epidemiol* (1965), 25(3): 609-610.
5. Benowitz NL. "Cotinine as a biomarker of ETS." *Epidemiol. Rev.* (1996), 18(2):188-204.
6. Reported by Samet JM and Wang SS. "Environmental Tobacco Smoke" Ch. 30 in IAQ Handbook, Edited by Spenger JD et al McGraw Hill (2001).
7. Taylor AE, Johnson DC, Kazemi H. "ETS and cardiovascular disease: A position paper from the CCPCC, Amer. Heart Assoc. - *Circulation* (1992), 86(2): 1-4.
8. Wallace CA. "The TEAM study: summary and analysis." EPA 600/6-87/0029. NTIS PB 88-100060. Washington, DC., U.S. EPA.
9. Whitmore RW et al. "Non-occupational exposure to pesticides for residents of two U.S. cities." *Arch. Environ. Contam. Toxicol.*, (1994) 26:47-59.
10. Leidy RB, Wright CG, Dupree HE. "Exposure levels to indoor pesticides," in *Pesticides in Urban Environments* pp. 283-296. ACS Symposium Series 5222 (1993). Amer. Chem. Soc., Wash, DC.
11. WHO. "Addressing the links between Indoor Air Pollution, Household Energy and Human Health." Report from WHO-USAID Global Consultation, Washington, DC 3-4 May, 2000.
12. Cooper KP and Alberti RR. "Effect of Kerosene heater emissions on IAQ and pulmonary function." *Ann. Rev. Resp. Dis.* (1984).
13. Rudel R. "Polycyclic Aromatic Hydrocarbons, Phthalates and Phenols." Ch. 34 in IAQ Handbook. Edited by Spenger JD et al. (2001) McGraw Hill.
14. Samet JM. "Radon." ibid. Ch. 40
15. Vallarino J. "Fibers." ibid. Ch. 37.

Chapter 4, Ten Consequences You can't Afford

1. Orent W. "Worrying about killer flu.' *Discover* (2005), 20 (2): 44-49.
2. National Institute of Allergy and Infectious Diseases (NIAID). Research on SARS, NIAID fact sheet. www.niaid.nih.gov .
3. Reuters, "SARS spread through airborne droplets." April 21, 2004 reported on MSNBC.com.
4. CDC, US Dept of Health and Human Services. "Recently asked Questions about SARS". www.cdc.gov.
5. WHO Report: "Indoor Air Quality Research." (1984). Copenhagen: WHO Regional Office for Europe.
6. WHO Report: "Indoor Air Pollutants, Exposure and Health Effects Assessment," (1982). Copenhagen. WHO Regional Office for Europe.

7. Mendell MJ. "Non-specific Symptoms in Office Workers: A Review and Summary of the Epidemiologic Literature." *Indoor Air* (1993), 3(4): 227-236.
8. Miller CS and Ashford NA. "Multiple Chemical Intolerance and IAQ." Ch. 21 in IAQ Handbook. Edited by Spenger JD et al. (2001) McGraw Hill.
9. Cullen M. "The worker with MCS." An overview in Workers with MCS: Occupational Medicine: State of the Art Review, Cullen M. (Ed), Philadelphia, Hanley & Belfus (1987), 2(4): 622-655.
10. Kreutzer R, Neutra R, Lashuay N. "Prevalence of people reporting sensitivities to chemicals in population based survey." *Amer. Jour. Epidemiol* (1999), 150(1): 1-12.
11. Wallace LC et al. "Association of personal and workplace characteristics with health, comfort and order: A survey of 3948 office workers in the building." *Indoor Air* (1993), 3:193-205.
12. Wallace LA, U.S. EPA. "Assessing Human Exposure to VOC" Ch. 33 in IAQ Handbook, Edited by Spengler JD et al (2001) McGraw Hill.
13. Krause C et al. "Occurrence of VOC in the air of 500 homes in the Fed Rep of Germany." *Proc. 4th Int. Conf. IAQ and Climate* (1987) 1, 102-106.
14. Godish R. "Aldehydes." Ch. 32 in IAQ Handbook. Edited by: Spengler JD et al. (2001) McGraw Hill.
15. International Agency for Research on Cancer. 1987. *IARC Monographs on the Evaluation of Carcinogenic Risk to Humans Overall. Evaluations of Carcinogenicity: An updating of IARC Monographs* Vols. 1 to 42, Suppl. 7, PP211-215.

Chapter 5, The Ozone Paradox

1. "A Brief History of Ozone.' www.ozonelab.com
2. Franken S. "The Application of Ozone Technology for Public Health and Industry." Unpublished, http://www.k-state.edu. 10/05
3. Referenced in www.yourairknowledge.com. July 1, 2005
4. "Epidemiologic Investigation to Identify Chronic Health Effects of Ambient Air Pollutants in Southern Calif." (2004). Principal Investigator, Dr. John Peters. CARB Contract No. 94-331.
5. Chiang C, Tasi C, Lin S, Hno C, Lo KV. "Disinfection of hospital waste water by continuous ozonization." *J. Environ Sci and Health* (2003), 38: 2895-2908.
6. Hawelaar AH et al. "Balancing the risks and benefits of drinking water disinfection: Disability adjusted life-years on the scale." Environ Health Persp (2000), 108(4): 315-321.
7. Mork DD. "Removing sulfide with Ozone." *Water Contamination and Purification* (1993): 34-37.
8. Kevin JG, Yousef AE, Chrism GW. "Use of ozone to inactivate microorganisms on lettuce." *J. Food Safety* (1999), 19: 17-33.
9. Zdrojewski F. "Ozone/UV water treatment: An alternative to chlorine treatment of tempering water. *Milling Journal* (2001): 40-42
10. Boeniger MF. "Use of ozone generating devices to improve IAQ." *Amer. Ind. Hygiene Assoc. Jour.* (1995), 56(6): 590-598.
11. Marsden JL. "Ozone used for Air Purification." (2005), Unpublished paper. www.k-state.edu.

Chapter 6, Eliminate Your Sources

1. U.S. EPA "Residential air cleaning devices: a summary of available information." (1990).
2. CDC Annual smoking-attributable mortality, years of potential life lost, and economic costs - US, 1995-1999. *MMWR Morb Mort Wkly Rep.* (2002), 51: 300-303
3. American Cancer Society: Guide for Quitting Smoking. www.cancer.org.
4. Hayden ACS. Fireplaces: Studies in Contrasts. www.hearth.com.
5. Adapted from Alexander M. Fall furnace maintenance, www.thisoldhouse.com.

6. American Lung Association, “Duct cleaning.” Sept. 2000. www.lungusa.org.
7. “Duct Cleaning: The Inside Dirt.” Better Homes and Gardesn, April 1999.
8. Referred to on http://departments.oxy.edu, Urban and Environmental Policy Institute (UEPI)
9. Referred to on www.non-toxic.info.

Chapter 7, Ventilate Your Space

1. Janssen JE. "The History of Ventilation and Temperature Control." *ASHRAE Journal* (1999), Sept., 47- 52
2. Addington DM. "The History and Future of Ventilation" Chap. 2 in the IAQ Handbook, Edited by Spengler, JD et al (2001) McGraw Hill.
3. Spaul WA. "Building-related factors to consider in IAQ evaluations." *Journal of Allergy Clin Immunol* (1994), 2(2): 385-389.
4. Saltzman A and Silberner J. "When each day is a sick day: How to get a breath of fresh air in a polluted office." *US News and World Report* (1989), March 13, 65-67.
5. *The Florida Times - Union*. "What's ailing you? Air quality problems in schools." Aug. 29, 2000.
6. Liddament M. "Ventilation Strategies" Chap. 13 in the IAQ Handbook, Edited by Spengler, JD et al (2001) McGraw Hill.
7. Holzman D. "Elusive culprits in workplace ills." *Insight* (1989), June 26, 44-45

Chapter 8, Filter The Air

1. McDonald B and Ouyang M. "Air cleaning - Particles." Ch. 9 in IAQ Handbook, Edited by Spengler JD et al., (2001) McGraw Hill.
2. Spengler JD, Harvard School of Public Health. Quoted from www.lungusa.org.
3. Underhill D. "Removal of Gases and Vapors." Ch 10 in IAQ Handbook, Edited by Spengler JD et al., (2001) McGraw Hill.
4. Nardell EA. "Disinfecting Air." *ibid.*, Chap. 11.
5. Adapted from Hames F. "The Seven Sins of Air Filter Manufacturers." *Indoor Air connections*. Referenced on www.allergybuyersclub.com.

Chapter 9, Purify The Air

1. Kowalski WJ, Bahnfleth WP, Whittam TS. "Bactericidal Effects of high airborne ozone concentrations on *E. coli* and *S. aureus*." *Ozone Science and Engineering* (1998), 20: 205-221.
2. Huebner RC. "Third party evaluation of the ability of 0.05 ppm ozone to inactivate common bacteria and fungi." Midwest Research Institute. Project No. 310413.1.001.
3. Marsden JL. "Ozone Used for Air Purification." (2005), Unpublished results available on www.k-state.edu.
4. Sherwood J. "Agro-terrorism preparedness." FDCH Congressional Testimony. U.S. Senate Agriculture Committee (July, 2005).
5. Purofirst "*Ozone*. 411 Information Please: Technical Data for fire, smoke and water damage restoration and reconstruction." (2000) p. 8.
6. U.S. EPA, "Ozone generators that are sold as air cleaners: An assessment of effectiveness and health consequences." www.epa.gov/iaq/pubs/ozonegen.html.
7. Kim JG, Yousef AE, Chrism GW. "Use of ozone to inactivate microorganisms on lettuce." *Journal of Food Safety* (1999), 19:17-33.
8. Pope DH, Erchler LW, Coates TF, Kramer JF, Soracco RJ. "The effect of ozone on *Legionella pneumophila* and other bacterial populations in cooling towers." *Current Microbiology* (1984), 10(2): 89-94.

9. Tenenbaum DJ. "The cleanroom: How clean?" Environmental Health Perspectives (2003), 111(5): 282-283.
10. Daniels SL. "Applications of Air Ionization for Control of VOC's and PM." Advances in, and Evaluation of, IAQ Control. Paper #918 (Session AB-7a). (2001).
11. Mitchell BW. "Effect of negative air ionization on airborne transmission of Newcastle Disease virus." *Avian Disease* (1994), 38:725-732.
12. Gabbay J. "Effect of Ionization on microbial air pollution in the dental clinic." *Environ. Res.* (1990), 52(1), :99.
13. Makela P, "Studies on the effects of ionization on bacterial aerosols in a burns and plastics surgery unit." *J. Hyg.* (1979), 83: 199-206.
14. Happ JW et al. "Effects of Air ions on submicron T1 bacteriophage aerosols." *Appl. Microb.* (1966), 14: 888-891.
15. Lee BU, Yermakor M and Grinshpun SA. "Removal of fine and ultrafine particles from Indoor Air Environment by the Unipolar Ion Emission." *Atmosph. Environment* (2004), 38: 4815-4823.
16. Grinshpun SA et al. "Evaluation of Ionic Air Purifiers for Reducing Aerosol Exposure in Confined Indoor Spaces." *Indoor Air* (2005), 15: 235-245.
17. Grinshpun SA et al. "Effect of Wearable Ionizers on the Concentration of Respirable Airborne Particles and Microorganisms." *Proc. Of the European Aerosol Conference,* Lupzig, Germany, Sept. 3-7, 2001; *Journal of Aerosol Science* (2001), 32 (Suppl. 1): 335-336.
18. Referenced on www.inspiringliving.com.
19. Nardell EA. "Disinfecting Air" Ch. 11 in IAQ Handbook. Edit by Spengler JD et al. (2001) McGraw Hill.
20. Kowalski WJ. "Design and Optimization of UVGI Air disinfection Systems." PhD Thesis, Penn. State Univ. (2001).
21. Xu, P et al. "Efficacy of UVGI of Upper-Room Air in Inactivating Airborne Bacterial Spores and Mycobacteria in full scale studies." *Atmospheric Environment* (2003), 37:405-419.
22. "The Application of UVGI to Control Transmission of Airborne Disease: Bioterrorism counter measure." *Public Health Reports,* Mar-Apr. 2003.
23. Kowalski WJ and Bahnfleth WP. "UVGI Design Basics for Air and Surface Disinfection." *Heating/Piping/Air Conditioning* (HPAC Engineering) Jan. 2000, p. 100-110.
24. Menzies D, Popa J, Hanley JA, Rand T, Milton DK. "Effect of ultraviolet germicidal lights installed in office ventilation systems on workers' health and wellbeing: double-blind multiple crossover trial." *The Lancet* (2003), 362 (9398): 1785-1791.
25. Kawalski WJ. Quoted on ww.cnn.com. Nov. 28, 2003.

Chapter 10, Space Age Technology

1. www.engr.psu.edu
2. Malik T. "Cleaning the Air: Smaller Detectors needed for Moon, Mars Missions." www.space.com.
3. www.spaceresearch.nasa.gov.
4. www.wcsar.engr.wisc.edu.
5. Fujishima A and Honda K. *Nature* (1972), 238: 37.
6. Ollis DF and El-Aleabi H, 9Eds.), Photocatalytic Purification and Treatment of Water and Air, Elsevier (1997), Publ., Amsterdam.
7. Mills A and Lahlunte S, *J. Photochem. Photobiol. A: Chem.* (1997), 108:1
8. Legrini O, Olivers E, Braun AM, Chems. Rev. (1993), 93: 671.
9. Huang Z et al. *J. Photochem. Photobiol A. Chem.* (1999).
10. Quotations from www.breathemoreeasily.com.
11. Falaras P et al. "Enhanced activity of silver modified thin film TiO2 photocatalysis." *Intl. Jour. Photoenergy* (2003), 5 (3): 123-130.
12. Bukhtiyarov VI. "Atomic Oxygen Species Adsorbed on Silver." Boreshov Institute of Catalysis, Laurentieva prosp., S, Novosibirsk, 6300%, Russia.
13. Schlogl R. "The current status of catalysis by silver." Fritz-Haber-Institut der Max-Plank-Ges.,

Faradayweg. 4-6, D-14195 Berlin, Germany.
14. Kang Yong Song, Young Tae Kwon, Guang Jin Choi, Wan In Lee. "Photocatalytic Activity of Cu/TiO2 with Oxidation State of Surface-loaded Copper." *Bull. Korean Chem. Soc.* (1999), 20(8): 957-960.
15. Linkous CA and Slattery DK. "Solar Photocatalytic Hydrogen Production from Water, Using a Dual Bed Photosystem." (2001). DOE Hydrogen Program Review. FL. Solar Energy Center, Univ. of Central FL.
16. Marsden JL. Food Science Institute, Kansas State Univ. (2005), unpublished results. www.ks-state.edu.

Chapter 11, Defend Your PERSONAL Space

1. Lee BU, Yermakov M, Grinshpun SA. "Removal of Fine and Ultrafine Particles from Indoor Air Environments by the Unipolar Ion Emission." *Atmospheric Environment* (2004), 38:4815-4823.
2. Lee BU, Yermakov M, Grinshpun SA. "Unipolar Ion Emission Enhances Respiratory Protection Against Fine and Ultrafine Particles." *Journal of Aerosol Science*, (2004), 35(11):1359-1368.
3. Grinshpun SA, Mainelis G, Trunov MA, Adhikari A, Reponen T, Willeke K. "Evaluation of Ionic Air Purifiers for Reducing Aerosol Exposure in Confined Indoor Spaces." *Indoor Air* (2005), 15: 235-245.
4. Lee BU, Yermakov M, and Grinshpun, SA "Filtering Efficiency of N95- and R95- Type Facepiece Respirators, Dust-Mist Facepiece Respirators and Surgical Masks, Operating in Unipolarly Ionized Indoor Air Environments." *International Journal for Aerosol and Air Quality Research* (2005), 5(3): 27-40.
5. Grinshpun SA, Mainelis G, Reponen T, Willeke K, Trunov MA and Adhikari A. "Effect of Wearable Ionizers on the Concentration of Respirable Airborne Particles and Microorganisms." *Proceedings of the European Aerosol Conference* (Leipzig, Germany, September 3-7, 2001), *Journal of Aerosol Science* (2001), 32 (Suppl.1):335-336.
6. Grinshpun SA, Adhikari A, Lee BU, Trunov M, Mainelis G, Yermakov M and Reponen T. "Indoor Air Pollution Control through Ionization." IN: Air Pollution: Modeling, Monitoring and Management of Air Pollution (2004) (Edit by Brebbia CA), WIT Press, Southampton, UK pp. 689-704.
7. Grinshpun SA, Lee BU, Yermakov M and McKay R. "How to increase the protection factor provided by existing facepiece respirators against airborne viruses: A Novel Approach." *Proceedings of the European Aerosol Conference* (2004) (Budapest, Hungary Sept 5-10, 2004), *Journal of Aeorsol Science*, EAC issue, 2:1263-1264.
8. Lee BU, Yermakov M, and Grinshpun SA. "Evaluation of the Ionic Air Purification Efficieny by the Electrical Low Pressure Impactor (ELPI)." *Program Addendum to the Annual Meeting of the American Association for Aerosol Research* - Late Breakings (Anaheim, Calif, Oct 20-24, '03) (2003), LB20
9. Grinshpun SA, Lee BU and Yermakov M. "Indoor Air Purification by Ionic Emission," *Abstracts of the American Industrial Hygiene Conference and Exposition* (Atlanta, GA, May 8-13, '04), #71, 18.
10. Grinshpun SA and Reponen T. "Respiratory Protection Against Airborne Biological Agents: A novel approach to enhance the efficiency of facepiece respirators. "*Abstracts of the Special Symposium "Perspectives on Biodefense: Science, Politics and Practice*" (Cincinnati, Ohio, Oct. 29-30 '04) (2004), A13.
11. Grinshpun, SA. "Respirator Performance with Infectious Agents. "*Abstracts of the CDC Workshop on Respiratory Protection for Airborne Infectious Agents."* GA, Nov 30-Dec. 1 '04) (2004), 3.1.
12. Dennis DP. "Treatment of Allergic Fungal Sinusitis by Decreasing the Environmental Fungal Load." *Archives of Environmental Health* (2003), 58(7): 433-441.
13. Schoeder I, Spira A, Madrid RE. "The Air Supply™ Personal Air Purifier and its Effect on Airborne Bacteria." UCLA Department of Microbiology and Molecular Genetics, LA, CA.
14. Gornostaeva O, Garate E and Pecota D. "Report on Testing the Bactericidal Effectiveness of

Sanitizer #2 and the Wand Sanitizer." (1999) Unpublished results. Report available, Univ. of California, Irvine.
15. Sherris D, Kern E and Ponikau J. "The Diagnosis and Incidence of Allergic Fungal Sinusitis." *Mayo Clinic Proceedings*. September 1999, p. 877.
16. Terman M and Terman JS. "Treatment of S.A.D. with a high-output negative ionizer." *J. Altern. Complem. Med.* (1995), 1(1):87-92.
17. Terman M and Terman JS. "A controlled trial of timed bright light and negative air ionization for treatment of winter depression." *Arch Gen Psychiatry* (1998), 55(10):875-82.

Chapter 13, Protect Your TRAVELING Space

1. "*In-Car Air Pollution: The Hidden Threat to Automobile Drivers*." International Center for Technology Assessment, Report No. 4: An Assessment of the Air Quality inside automobile passenger compartments, ICTA Washington, DC, July 2000.
2. Dockery D et al. "An Association between Air Pollution and Mortality in Six Cities." *New Engl. J. Med.* (1993), 329:1753.
3. Pope CA et al. "Particle Air Pollution as a Predictor of Mortality in a Prospective Study of U.S. Adults." *Amer J. of Resp and Crit. Care Med.* (1995) 151:669.
4. Kaiser J, "Panel Scores EPA on Clean Air Science." *Science* (1998) 280: 193-194
5. Ptak TJ and Fatlon SL. "Particulate Concentration in Automobile Passenger Compartments." *Particulate Science and Technology* (1994), 12:313-322.
6. Chan CC et al. "Driver Exposure to VOC's, CO, Ozone and NO2 Under Different Driving Conditions." *Environmental Sciences Technology* (1991), 25:964-972.
7. Chan CC et al. "Commuter Exposure to VOC's in Boston, Mass." *Journal of Air & Waste Mgt. Assoc.* (1991), 41:1594-1600.
8. California Air Resources Board Study (1998), www.arb.ca.gov.
9. "Perspectives in Disease Prevention and Health promotion: CO - A preventable Environmental Health Hazard." *Morbidity and Mortality Weekly Report 31* (Oct 9, 1982): 529, Dept of Transportation, NHTSA.
10. Petersen WB and Allen R. "CO Exposures to LA Area Commuters." *J. Air Pollution Control Assoc.* (1982), 32: 826-833.
11. Konshki PA et al. "Vehicle Occupant Exposure to CO." *Journal of Air & Waste Mgt. Assoc.* (1992), 42:1603-1608.
12. Conor Dougherty, "Products that promise to keep you healthy on planes." *The Wall Street Journal*, Jan. 07, 2006.
13. VanNetlen C. "Air Quality and Health Effects associated with the operation of BAe 146-200 air craft." *Appl. Occup. Environ Hyg.* (1998), 13:733-739.
14. "Airborne Infections on Commercial Flights." Report 10 of the Council on Scientific Affairs (A-98), *American Medical Association.* www.ama-assn.org.
15. Ko G, Thompson KM and Nardell EA. "Estimation of Tubercolosis Risk on a Commercial Airline." *Risk Analysis* (2004), 24(2): 379-387.
16. Wenzel R. "Airline Travel and Infection." *N. Eng. J. Med.* (1996), 334(15):981-982.
17. *USA Today*, July 10, 2000. "Davis R Group urges mandatory air filters on jets."
18. Hocking MB. "IAQ: Recommendations relevant to aircraft passenger cabins." *AIHA Journal* (1998), 59:446-454.
19. Phillips EH. "FAA mandates lower CO2 levels." *Aviation Week and Space Technology* (1997), 146(3):44.
20. Adapted from "Overview of IAQ Problems in Airplanes." *www.aerias.org*.
21. Grinshpun SA et al. "Evaluation of ionic air purifiers for reducing aerosol exposure in confined indoor spaces." *Indoor Air* (2005), 15: 235-245.
22. Bounds W. "The neurotic traveler can rely on an armory against their fears." *The Wall Street Journal*, Nov. 26, 1999.

Chapter 14, Clean Up Your WORK Space

1. Chelsea Group (July 31, 2000). "People are willing to spend money to improve IAQ." www.safetyonline.com.
2. Mendel MJ. "Non specific Symptoms in office workers: A Review and Summary of the Epidemiologic Literature." *Indoor Air* (1993), 3(4): 227-236.
3. Fish WJ, Mendell MJ, Daisey JM, Faulkner D, Hodgson AT, Nematallaki M and Macker JM. "Phase 1 of the Calif. Health Building Study: A Summary." *Indoor Air* (1993), (4): 246-254.
4. TenBrinke J et al. "Development of New VOC exposure metrics and their relationship to SRS symptoms." *Indoor Air* (1998), 8:140-152.
5. Wallace R, Wilcox TG and Sieber WK. "Workplace Characteristics associated with Health and Comfort Concerns in Three Office Buildings in Washington, DC." *IAQ '91 - Healthy Buildings* (1991), ASHRAE, Atlanta.
6. Malkin R, Wilcox TG and Sieber WK. "The NIOSH Indoor Environmental Evaluation Experience, Part Two: Symptom Prevalence." *Applied Occupational and Environmental Hygiene* (1996), 11(6):540-545.
7. Wamble SE, Girman JR, Ronca EL, Axelrod R, Brightman HF and McCarthy JF. "Developing Baseline Info on Buildings and IAQ (BASE '94), Part 1: Study Design, Building Selection, and Building Descriptions." *Healthy Buildings '95, An Intl. Conf on Healthy Building in Mild Climate* (1995), Vol. 3:1305-1310.
8. Adapted from Kotelchuk, David. "IAQ: An Old Problem Reappears," *UE News*, Mar/Apr. 2001 as published on www.ranknfile-ue.org.

Chapter 15, Influence Your COMMERCIAL Space

1. Miller A. "Bad air breeds ailments in homes, schools, offices." A Special Report. A*tlantic Journal Constitution* (July 2003).
2. Rynders D. "Inside schools: The air we breathe." *CEC Environmental Exchange Newsletter* (2004)
3. "Schools seek cleaner, fresher air." *Associated Press*, Aug. 15, 2003 on www.MSNBC.com, Mar. 4, 2004.
4. Miller A. "New building practices can clean indoor environments." A Special Report. *Atlantic Journal Constitution* (July 2003).
5. Ball TM, Castro-Rodriguez JA, Griffith KA, Holberg CJ. "Siblings, daycare attendance, and the risk of asthma and wheezing during childhood." *New Eng. J. Med* (2000), 343(8):538-543.
6. Ruotsalainen R, Jaakkola N, Jaakkola JK. "Dampness and molds in daycare centers as an occupational health problem." *Int Arch Occup Environ Health* (1995), 66:369-374.
7. PerNafstad, Hagen JA Oie L, PerMagnus J, Jaakkola JK. "Daycare centers and Respiratory Health." Pediatrics (1999), 103:753-758.
8. Commercial Buildings Energy Consumption Survey (1999) Energy Information Administration, Washington, DC www.eia.doc.gov.
9. IAQ and You. *Lodging Hospitality*. Oct. 1999, 50-53.
10. Thompson B. "Travel - The dish on hotel air." *The Wall Street Journal*, June 18, 1999, p.W14.
11. U.S. Newswire Release, Apr. 19, 2005. "Pilot Project Enhances IAQ in Hotel Rooms; Chicago firm's Concept Targets Allergy, Asthma Sufferers."
12. Gould KL. "Biofiltration could become an effective means of combating poor IAQ." *Architectural Record* (1999), 187(10):214.
13. American News Service. "Breath of fresh air in tourism." *The Times Union* (Albany, NY) Aug. 13, 2000; A1b.

INDEX